Bulletin 51

DEPARTMENT OF THE INT

BUREAU OF MIN

JOSEPH A. HOLMES, Director

I0043285

THE ANALYSIS OF
BLACK POWDER AND DYNAMITE

BY

WALTER O. SNELLING AND C. G. STORM

WASHINGTON
GOVERNMENT PRINTING OFFICE
1913

First edition. March, 1913.

CONTENTS.

ILLUSTRATIONS.

THE ANALYSIS OF BLACK POWDER AND DYNAMITE.

By Walter O. Snelling and C. G. Storm.

INTRODUCTION.

Although descriptions of the methods of analysis of explosives are to be found in many books on explosives, and in works on engineering chemistry or chemical analysis, most of these descriptions are incomplete and lacking in details. The methods of analysis employed in the laboratories of most explosives factories are frequently treated as trade secrets, and very little information is published from such laboratories.

This bulletin outlines the methods of analysis that are used by the Bureau of Mines in the examination of certain classes of explosives. The present form of most of these methods has been worked out in the bureau's explosives laboratory. The methods employed by Prof. C. E. Munroe were taken as a basis, and were elaborated to meet the demands incident to the treatment of complicated mixtures and to the development of the explosives art. A subsequent bulletin will discuss the methods of analysis of "permissible" explosives, many of the latter being of decidedly complicated character and requiring special treatment. This bulletin presents the methods of analysis of "ordinary" dynamite, and the ammonia, gelatin, low-freezing, and granular dynamites, and the common grades of black gunpowder and black blasting powder. The bulletin is published by the bureau for the information of all persons interested in explosives and their safe and efficient use in mining work.

As the term "ordinary" dynamite, though much used, has no conventional meaning, and may be used to cover a wide variety of compositions of matter, it may be noted that the standard dynamite used at the Pittsburgh testing station is a good example of the "ordinary" dynamite known in this country. This testing station dynamite has the following composition:

Composition of Pittsburgh testing station dynamite.

	Per cent.
Nitroglycerin	40
Sodium nitrate	44
Wood pulp	15
Calcium carbonate	1

As most permissible explosives contain only the constituents found generally in the various types of ordinary dynamite, the chemist will usually find it possible to analyze such explosives either wholly or partly by following the general methods of analysis here given for the type of explosive that seems most closely related to the one under examination. The methods of extraction with ether, with water, etc., here outlined are general methods which are applied with equal success to all classes of explosives, and therefore by the use of these general methods, following a thorough qualitative examination, little difficulty should be met except with those classes of permissible explosives that contain large amounts of salts holding water of crystallization, such as alum and magnesium sulphate, or those containing an unusual number of uncommon constituents. Even with such explosives, however, if the information desired is principally in regard to the percentages of explosive ingredients (nitroglycerin, ammonium nitrate, etc.), the methods outlined in this bulletin may be satisfactorily followed.

DYNAMITE.

"Ordinary" dynamite consists essentially of nitroglycerin absorbed in some porous material. Owing to its physical condition and its extreme sensitiveness to shock, liquid nitroglycerin is not suitable for use as an explosive in mining and quarrying, but when nitroglycerin is absorbed in a porous material a more or less plastic mass is obtained which is far less sensitive to shock than liquid nitroglycerin, although, when properly fired by means of a detonator, it retains most of the explosive properties of nitroglycerin. Among the many substances that have been used as absorbents for nitroglycerin are sawdust, wood pulp, ground mica, and infusorial earth (kieselguhr), or mixtures of these substances with alkaline nitrates and other substances.

It is usual to classify absorbents for nitroglycerin as active and inactive. Pulverized gunpowder, for example, or mixtures of wood pulp with sodium nitrate or other oxidizing agents, represent "active" absorbents, whereas mica, kieselguhr and similar materials, which play no part in the explosive reactions and which are employed merely to absorb or retain the liquid nitroglycerin, form the so-called "inactive" absorbents.

The type of dynamite most generally used to-day consists of nitroglycerin absorbed in a mixture of wood pulp and sodium nitrate, and to this mixture is usually added a small amount of some antacid such as calcium carbonate, magnesium carbonate, or zinc oxide. This antacid is added in the belief that it increases the stability of the resulting explosive by neutralizing such small amounts of free acid as may be produced by the decomposition of the nitroglycerin during long storage.

The analysis of dynamite is best carried out by first separating, with ether or some other appropriate solvent, the nitroglycerin from the dope in which it is absorbed. After the nitroglycerin has been thus removed, the soluble nitrate in the dope may be removed by dissolving in water; the antacid may then be dissolved in dilute acid, and the residue insoluble in ether, water, dilute acid, etc., may be directly determined by weight.

In its simplest form, therefore, the analysis of dynamite consists in the removal of the constituent materials, one by one, through the use of appropriate solvents. Dynamites of the most complicated composition may usually be analyzed in this way, through selective solution. In the present paper the methods of analyzing ordinary types of dynamite are discussed, and those that have been found best in an experience covering several thousand analyses are stated.

PHYSICAL EXAMINATION.

Upon receiving a sample of explosive for analysis it is desirable to record full information in regard to the size and weight of each cartridge, with a complete copy of any lettering that may appear on the wrapper. It is also advisable to record the nature of the outer wrapping paper (such as ordinary paper, parchmentized paper, or paper coated with paraffin), and whether the cartridge has been redipped; that is, placed in a paraffin bath after being filled. Whether a cartridge has been redipped can usually be determined by carefully opening the wrapper. If there is a greater thickness of paraffin near the edge where the sheet overlaps, or if the overlapping edge is attached to the adjacent portions of the paper by means of an adhering deposit of paraffin, it may be assumed that the cartridge has been redipped.

DETERMINATION OF GRAVIMETRIC DENSITY.

It is possible to determine approximately the gravimetric density or apparent specific gravity of a cartridge of explosive by measuring carefully the length and circumference of the cartridge, calculating from these figures the volume in cubic centimeters, and then dividing the weight in grams of the cartridge by this figure. However, experiments made at the bureau's explosives laboratory have shown that even with the most careful measurements the figures thus obtained are liable to be in error by as much as 10 to 20 per cent, a difference entirely too great to make the method permissible for exact work. With some redipped cartridges weighing in water has given satisfactory results, but cartridges seldom have a coating of paraffin so complete as to permit the use of this method. Accordingly a method was sought that would at all times give satisfactory results even with cartridges that had not been redipped.

The volume of the cartridge can be determined conveniently by using sand instead of water as the measuring material. A weighed glass cylinder about 30 cm. high and 5 cm. in inside diameter is filled with fine sand (preferably sea sand) that has been sifted through a 60-mesh sieve. A straight edge is drawn across the top of the cylinder, the level of the sand being left flush with the top edge, and the weight of the cylinder and contained sand is determined. From this weight the weight of the cylinder is subtracted and the result is the weight of the sand, which, divided by the weight of water required to fill the cylinder, gives the apparent specific gravity of the sand used. All the sand except enough to fill the cylinder to a depth of about 1 inch is now poured out, a weighed cartridge of the explosive is placed in the cylinder, and sand added until the cylinder is filled flush to the top as before, when it is struck with the straight edge and then the weight of cylinder and sand and cartridge is noted. From these figures the weight of sand displaced by the cartridge is found. This weight divided by the apparent specific gravity of the sand gives the volume of the cartridge. The weight of the cartridge divided by its volume gives its apparent specific gravity or gravimetric density. This determination leaves the cartridge in condition for use in sampling, if desired.

In making this determination care should be taken that the cylinder is filled each time in exactly the same manner, the sand being poured in slowly and not packed by jolting, shaking or otherwise. Repeated determinations of the weight of sand required to just fill the cylinder will prove that with proper care uniform results may be obtained; in practice this method has been found to be both rapid and exact.

TEST FOR LIABILITY OF EXUDATION.

To determine whether there is liability of leakage of nitroglycerin from cartridges containing this explosive, it is always advisable to make an exudation test, which indicates the amount of nitroglycerin that may be lost by the explosive tested under prescribed conditions. The tests most commonly used for this purpose are the 40° test, the pressure test, and the centrifugal test.

40° TEST FOR EXUDATION.

In the 40° test a cartridge of the explosive under examination is placed in a vertical position in an oven heated to 40° C. Some small perforations are made in the wrapper at the ends of the cartridge, and the cartridge is then placed on end on a small wire tripod in a small glass beaker or cylinder. The whole is then placed for six days in an oven maintained at a constant temperature of 40° C. At the end of this time an examination is made to see if any leakage of nitroglycerin in the form of drops has occurred. Such leakage may be taken as evidence that there is too much nitroglycerin in the explosive for the amount of absorbent material present, or that the dope used is deficient in absorbing capacity.

PRESSURE TEST FOR EXUDATION.

Before the centrifugal test was employed, it was customary to use a pressure test for exudation which consisted in exposing a sample of the dynamite to a definite pressure produced by a weight on a lever arm, and determining the amount of nitroglycerin thereby forced out of the dynamite. Many modifications of this test have been tried, in which cotton, blotting paper, or other absorbent material have been used to hold the nitroglycerin forced out from the explosive. The pressure test is unreliable and hence is not satisfactory in use. For example, by this test an explosive that contains a certain amount of a mixture of sawdust and wood pulp shows less exudation than one in which the same amount of absorbent is used in the form of wood pulp alone, a result that is clearly incorrect, since the absorbing power of wood pulp is much greater than that of sawdust. The reason for the more favorable result when the absorbent contains sawdust is that the particles of sawdust are packed together to form a cellular mass, which incloses the particles of wood pulp holding the nitroglycerin, thereby in a great measure protecting them from pressure.

CENTRIFUGAL TEST FOR EXUDATION.

The use of centrifugal force as a means of measuring the completeness with which nitroglycerin is absorbed in an explosive was some years ago suggested to Col. B. W. Dunn, chief inspector of the bureau for the safe transportation of explosives, by T. J. Wrampelmeier, an inspector in that bureau, and a device for testing the value of this method was made use of by C. P. Beistle,[a] chief chemist of that bureau. The method employed was to place the explosive, together with a perforated disk of vulcanite and an absorbent material such as cotton, within a glass tube, the tube being then placed in a centrifuge. The increase in weight of the cotton after rotation was taken as a measure of the amount of nitroglycerin lost by the explosive during the process. This apparatus gives much more satisfactory results than the former methods of testing by pressure alone, but owing to the fact that the cotton becomes compressed during rotation, thus changing the position of the vulcanite disk, the apparatus at

---

[a] Report of chief inspector of the bureau for the safe transportation of explosives, February, 1909.

times gives discordant results, and the figures from tests of any two explosives are not proportional to the relative tendency toward leakage of nitroglycerin under normal conditions.

One of the authors designed a centrifuge attachment which is now being used with reliable results by the Bureau of Mines. This apparatus is shown in Plate I, A. Two samples of the explosive are placed in ordinary porcelain Gooch crucibles, without mat, the crucibles being held above two other nonperforated crucibles in the manner shown. A small amount of cotton is placed in each of the lower crucibles to receive the exuded nitroglycerin, and the loss by exudation is determined by weighing the crucibles containing the explosive before and after rotation. The circle of rotation made by the bottom of the crucibles is 14 cm. in diameter, and the standard velocity of rotation is 600 revolutions (30 turns of the handle) per minute. The usual test consists in placing 8 grams of explosive in each of the upper crucibles, and determining the loss in weight after rotating at the velocity of 600 revolutions per minute for 5 minutes at a temperature of about 20° C. If the explosive does not lose more than 5 per cent in weight it is considered to have satisfactory absorbing capacity, but if more than 5 per cent is lost its absorbent properties are considered deficient, and in the transportation or use of such an explosive there is considered to be liability of accident.

TEST FOR STABILITY.

Many tests have been proposed for determining the stability of explosives under the influence of heat, and much has been written in regard to the comparative accuracy of these different tests. This field is now being investigated by the bureau, and a special report thereon will be issued. At present all mining explosives examined by the bureau are tested for stability by means of the Abel heat test.

ABEL TEST.

The Abel stability test depends upon the fact that when potassium iodide is decomposed in the presence of starch, the iodine liberated reacts with the starch to form a colored body. The explosive to be tested is placed in a stoppered test tube and heated in a constant-temperature bath until the oxides of nitrogen liberated as decomposition products make a brownish color on a strip of potassium iodide starch paper suspended in the tube above the explosive. The stability of the explosive is judged by the time required for the production of a coloration of a standard intensity. The apparatus used in the Abel test is illustrated in Plate I, A.

Two grams of explosive in its original condition, without preliminary drying or preparation other than thorough mixing, is placed in

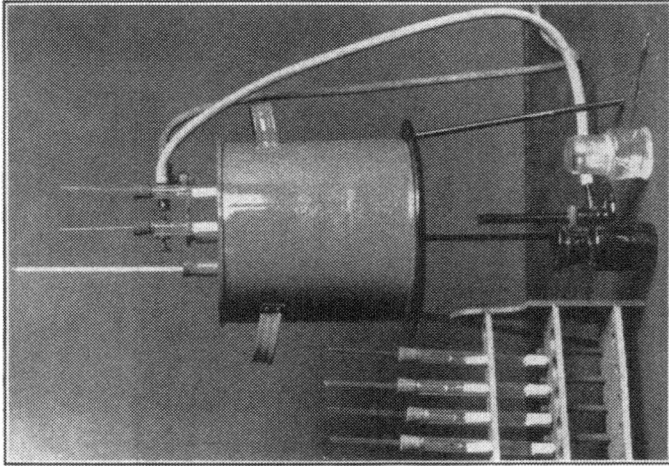

B. APPARATUS FOR ABEL HEAT TEST.

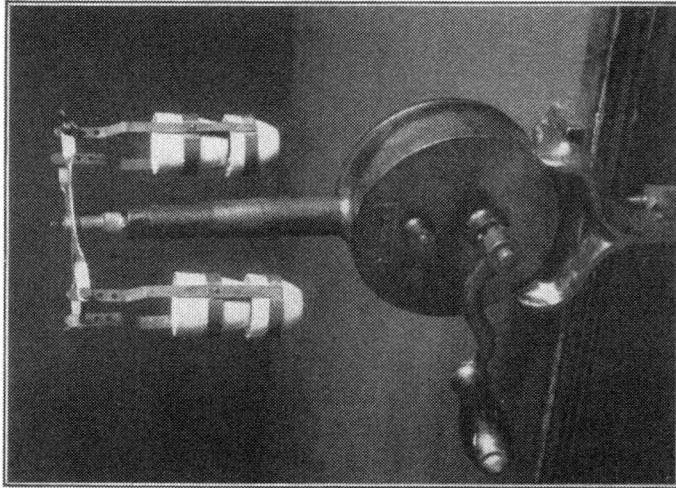

A. CENTRIFUGE FOR EXUDATION TEST OF DYNAMITE.

a glass tube. The tube is of standard dimensions as follows: Length, 14 cm. (5½ inches); inside diameter, not less than 1.27 cm. (½ inch); outside diameter, not more than 1.59 cm. (⅝ inch); thickness of the glass, about 1.2 mm. ($\frac{3}{64}$ inch). The tube is closed by a clean, tightly fitting cork stopper, through which passes a glass rod provided with a platinum hook, fused into the lower end for holding the test paper. The test papers, in pieces about 2.5 cm. (1 inch) by 1.0 cm. (⅜ inch), are hung on the platinum hooks (forceps being used in handling) and the upper half of each strip of test paper is moistened with a solution of equal volumes of pure glycerin and water.

The test paper used is potassium iodide starch paper, similar to that prepared by Eimer & Amend (Frankford Arsenal formula), or prepared by a standard method as described below. The heat-test bath is placed so that a good light—not direct sunlight—is transmitted through the test papers to the operator. The bath temperature is maintained constant within 0.5° C. of the desired temperature (71° C.), the thermometer being so immersed that the bottom of the bulb is 2½ inches below the top of the bath.

All determinations are made in duplicate. The shorter time required to bring one of the two test papers to the prescribed tint determines the test of the explosive, except in case of wide variation in results, when two or more additional samples are tested.

After each test the cork stopper of each tube is either discarded or carefully washed and dried and in any case is frequently renewed. The tube and the rod are carefully cleaned after each test, ether or other solvent being used to remove nitroglycerin, etc. They are then washed with water, and finally rinsed with distilled water. All the parts of the apparatus are dried in a steam oven at 100° C. The apparatus is at all times protected from laboratory fumes.

The tube is inserted to a depth of 2½ inches below the top of the bath, the water in the bath being within one-fourth inch of the top. The time of placing the tube in the bath is recorded, and the test is considered completed on the appearance of a brownish line at the lower edge of the moist portion of the paper, of the same intensity as the line on a standard-tint paper prepared as described below. It should be noted, however, that the brownish color on the test paper may at times be spread over a considerable portion of the paper, not forming a sharp, well-defined line. The operator should judge what would be equivalent to the standard tint over a width of one-half to 1 mm.

Caramel standard tint paper.[a]—The tint paper used as a standard color comparator for the test is made as follows: A solution of caramel in distilled water is prepared, of such concentration that when diluted to 100 times its volume (10 c. c. diluted to 1 liter) the tint of

a 30th Ann. Report H. M. Inspector of Explosives, 1905, p. 236.

the solution equals that produced by the Nessler test on adding 2 c. c. of Nessler reagent to 100 c. c. of water containing 0.000075 gram of NH_3, or 0.0002305 gram of NH_4 Cl. With a fine brush or pen dipped into this caramel solution fine lines are drawn on strips of filter paper (Schleicher and Schüll, 597). These strips are cut to the size of the regular test papers (1 inch by $\frac{3}{8}$ inch) so that the brown line crosses each piece near the middle of its length. The line should be $\frac{1}{2}$ to 1 mm. in width when dry. A piece of this standard tint paper should be placed in an empty tube beside those being tested, so that a comparison of color may be made.

Preparation of test paper for Abel test.[a]—The paper used in preparing the test paper is Schleicher and Schüll's filter paper 597. This is cut in strips about 6 by 24 inches, and after being washed by immersing each strip in distilled water for a short time is hung up to dry overnight. The cords on which the paper is hung are clean and the room is free from fumes. The washed and dried paper is dipped in a solution prepared as follows:

The best quality of potassium iodide obtainable is recrystallized three times from hot absolute alcohol, dried, and 1 gram dissolved in 8 ounces of distilled water. Cornstarch is well washed by decantation with distilled water, dried at a low temperature, 3 grams rubbed into a paste with a little cold water, and poured into 8 ounces of boiling water in a flask. After being boiled gently for 10 minutes, the starch solution is cooled and mixed with the potassium iodide solution in a glass trough.

Each strip of filter paper is immersed in the above-mentioned mixture for about 10 seconds and is then hung over a clean cord to dry. The dipping is done in a dim light and the paper left overnight to dry in a perfectly dark room. Every precaution is taken to insure freedom from contamination in preparing the materials and from laboratory fumes that might cause decomposition. When dry the paper is cut into pieces about $\frac{3}{8}$ by 1 inch and is preserved in the dark in tight glass-stoppered bottles, the edges of the large strips being first trimmed off about one-fourth inch to remove portions that are sometimes slightly discolored. When properly prepared the finished paper is perfectly white, any discoloration indicating decomposition due to contamination.

SAMPLING.

The first step to be taken in the analysis of dynamite, as with any other material, is the careful preparation of a sample. Dynamite is offered in commerce in the form of cylindrical "sticks" or cartridges, usually three-fourths inch to $1\frac{1}{2}$ inches in diameter and 8

---

[a] Storm, C. G., Proc. 7th Inter. Congress Applied Chem. 1909; Jour. Ind. and Eng. Chem., vol. 1, 1909, p. 802.

inches long.   For special work dynamite is made in cartridges up to 5 inches in diameter, but owing to restrictions in railroad transportation the length of cartridges of dynamite does not vary greatly, and cartridges over 8 inches long are unusual.

A cartridge of dynamite consists of a c vering of paper, sometimes waxed or parchmentized and often coated with paraffin, within which the dynamite is more or less tightly packed to give it the density desired.   In the manufacture of dynamite the paper "shell" is first made, and is then packed with the explosive, after which the cartridge is sometimes "redipped," by which is meant that the cartridge is plunged into a bath of paraffin heated slightly above its melting point.   The paraffin closes any openings in the wrapper and tends to make the cartridge waterproof.   This operation of "redipping" is of interest in connection with the analysis of explosives chiefly because of the opportunity it affords for the entrance of paraffin into the explosive.   When a paper shell is not perfectly made some paraffin is apt to find its way through the paper shell and be absorbed by the wood pulp.   In sampling, care should always be taken to remove any paraffin that has found access through the ends of the cartridge, since obviously such paraffin is not to be considered as a normal constituent of the explosive, and care should also be taken in unwrapping the cartridge to prevent scales and flakes of paraffin from becoming mixed with the sample.

The best method of sampling consists in opening the wrapper of each cartridge, spreading it out and cutting off from 3 to 5 cm. from each end of the roll of explosive thus exposed.   These ends are rejected and the remainder carefully broken up to form a homogeneous mass.   When a sample representing a large quantity of powder—for example, a day's output of a factory—is to be prepared, a number of cartridges are taken from different mixings, the central portions of each cartridge are selected in the manner described, and these portions are finely broken up in a large porcelain evaporating dish or on a sheet of paraffined paper.   The mass is carefully stirred with a clean spatula, or is rolled from side to side upon the paraffined paper in the manner usually followed in preparing a sample of ore for assay.   The stirring of the sample, or its rolling back and forth upon the paraffined paper, should occupy not less than five minutes, and the best results are obtained from such sampling when the explosive has been previously broken up to as fine a meal as possible by crumbling in the fingers or by gentle pressure with the spatula. A spatula suitable for this purpose is made of horn, or of wood saturated with paraffin so that it will not absorb nitroglycerin.

From the large sample prepared as described a sample of 50 to 100 grams is taken by selecting small portions from different parts of the mixed material, mixing these portions, and placing the mixture in the sample bottle.

In preparing a sample of dynamite there are several factors that must be constantly borne in mind. If a thoroughly mixed sample is prepared and allowed to remain for some time in a sample bottle, a segregation occurs, and the lower portions of the material in the sample bottle become richer in nitroglycerin at the expense of the upper portions. In some dynamites, particularly one that contains almost as much nitroglycerin as its absorbent base can hold, this change occurs rapidly, and may make a difference of several per cent in a few days. The tendency to segregate is greatest in a tall bottle, and is favored by warmth, the action taking place several times as rapidly at 30° C. as at 20° C. An indication of the amount of segregation that is possible in dynamite is seen from the results of the following experiment:

A sample of dynamite was prepared by carefully breaking up and sampling two sticks of 60 per cent dynamite, about 300 grams being taken and placed in an ordinary bottle of 250 c. c. capacity (diameter about 6 cm., height about 14 cm.). At the time the sample was placed in the bottle the analysis of the material gave the following results:

Analysis of 60 per cent dynamite.

[W. C. Cope, analyst.]

	Per cent.
Moisture	1.40
Nitroglycerin	60.60
Potassium nitrate	18.64
Calcium carbonate	1.25
Wood pulp	18.11

At the end of 10 days' exposure to a temperature of 32° to 33° C. (average, 32.5° C.) samples were taken from the top, middle, and bottom of the material in the bottle, and the following results were obtained:

Analyses of different parts of exposed sample of 60 per cent dynamite.

[W. C. Cope, analyst.]

Constituents.	Sample from top.	Sample from middle.	Sample from bottom.
	Per cent.	Per cent.	Per cent.
Moisture	1.25	1.16	1.17
Nitroglycerin	59.46	60.55	62.86
Potassium nitrate	19.91	18.44	17.42
Calcium carbonate	1.37	1.33	1.26
Wood pulp	18.01	18.52	17.29

Another experiment was made to illustrate segregation in cartridges themselves.

Whole cartridges of 40 per cent, 45 per cent, and 60 per cent dynamite were placed upright (on end) in a constant-temperature oven at 40° for a period of four weeks, the cartridges being supported by small wire tripods within glass cylinders. Pinholes were punched in the bottom of the wrappers to allow the escape of any accumulation of nitroglycerin, but no such leakage occurred.

At the end of four weeks samples were taken from the top and bottom of each cartridge, each sample representing about a 2-inch length of cartridge. The following results of analyses of these samples show the segregation of the nitroglycerin:

Analyses of parts of samples of dynamite left standing on end for four weeks at 40° C.

[J. H. Hunter, analyst.]

Grade of dynamite, per cent	40	45	60
Part of cartridge sampled.	Nitroglycerin.		
	Per cent.	*Per cent.*	*Per cent.*
Top	38.20	45.07	57.27
Bottom	40.96	47.98	63.26

From the above described experiments it will be seen that the sampling of dynamite involves problems not met in the preparation of ore for assay.

To determine any possible segregation due to nitroglycerin adhering to the dish in which a sample is mixed, 500 grams of dynamite was mixed in a large evaporating dish, and, when thorough mixing had been effected, was poured on a piece of paraffined paper. An analysis was then made of the large sample upon the paraffined paper, and the material adhering to the sides of the evaporating dish was removed by means of ether and analyzed. It was found that there was little difference between the composition of the two samples. It has been suggested that considerable changes in the moisture content of dynamite can occur during sampling, but this opinion does not seem correct. In the factory dynamite is usually mixed in an open mixer exposed to the atmosphere; consequently the percentage of moisture taken up or lost in the short length of time that the explosive is exposed to the air during sampling should be proportional only to the difference between the hygrometric condition of the air at the time the dynamite was made and that at the time it was sampled. To investigate this question further an experiment was made as follows:

Two cartridges of one-half pound (226 grams) each were mixed together quickly, and with as little exposure to the air as possible. This original sample was found to contain 1.10 per cent moisture.

A 100-gram portion of this large sample was then mixed on a large watch glass for 10 minutes on a damp day when the humidity of the air as determined by hygrometer readings was 75 per cent. Another 100-gram portion was similarly treated on a dry day when the humidity was only 19 per cent. Determinations of moisture were then made on these two portions with the following results:

Effect of exposure during sampling on moisture content of dynamite.

Treatment of sample.	Moisture.
	Per cent.
Without undue exposure	1.10
Stirred 10 minutes in moist atmosphere	1.25
Stirred 10 minutes in dry atmosphere	1.10

From these results it will be seen that the condition of the atmosphere has little effect on the moisture content of a sample when the sampling is done with reasonable dispatch and when atmospheric conditions are not abnormal.

CHEMICAL EXAMINATION.

QUALITATIVE EXAMINATION.

When no information is available as to the class to which an explosive to be analyzed belongs, a complete qualitative analysis is desirable, so that the proper methods of quantitative separation may be followed.

For the qualitative examination of a dynamite of any type a sample of 20 to 25 grams is most convenient. The sample is placed in a 1-inch test tube, which is then filled to about two-thirds of its depth with ether, stoppered, and well shaken. The ether is decanted through a filter paper and fresh ether is added to the tube. This treatment is repeated several times, and the residue is finally transferred to the filter and again washed with fresh ether. After the ether is drained off, the filter paper with its contents is removed from the funnel, spread out on a glass plate, and placed in a drying oven for a short time until nearly all the ether has evaporated. The dry residue is transferred from the filter back to the test tube, and is then ready for treatment with cold water, to remove the water-soluble constituents.

The ether solution is evaporated on a steam bath or electric heater at a low temperature until all odor of ether has disappeared. If the evaporation has caused the deposition of water in the beaker with the extract, this water is removed by placing the beaker in a vacuum desiccator for an hour or two. The presence of nitroglycerin is readily noted, the characteristic oily appearance of nitroglycerin and its viscosity serving to identify it. A convenient chemical test is to mix a drop of the heavy liquid, supposed to be nitroglycerin, with 1 or 2 c. c. of concentrated sulphuric acid in a test tube, and then to add about 1 c. c. of mercury. No stopper of any sort should be placed in the tube, but it should be shaken quickly from side to side so as to cause intimate contact of the mercury with the acid mixture. For 1 or 2 minutes little effect will be observed, but after a short time there will be noted, if the material under examination is nitroglycerin, the evolution of bubbles of gas from the liquid at the point where it comes in contact with the mercury, and the characteristic smell of nitrogen oxides will be noted. The nitric oxide (NO) produced is colorless, but upon coming in contact with the air it turns red or reddish-brown, forming nitrogen peroxide (NO_2). A simpler test for the presence of nitroglycerin, though by no means so satisfactory as the one just described, consists in taking up a small amount of the liquid in a capillary glass tube, and holding it cautiously in a flame. A strong detonation will result if the liquid is nitroglycerin. It is of course evident that there should be not more than a tiny fraction of a drop of nitroglycerin used in such a test, the best results being obtained when a very thin and small capillary is used, containing not more than 0.01 of a drop of the liquid tested.

If sulphur is present in the ether extract it will crystalize in needles or small granular masses in the residue upon the evaporation of the ether. The crystals of sulphur can be removed, washed free from nitroglycerin with a little acetic acid (70 per cent or glacial), after which they should be washed with water, dried, and heated over a flame. The odor of sulphur dioxide (SO_2) will identify the crystals as sulphur. Trinitrotoluene appears as long yellowish needles, which may be recrystallized from alcohol and identified by their melting point (80° C.), or by the color test with potassium or sodium hydroxide.[a] This color test is valuable as a means of identifying many of the nitrosubstitution products, although in a mixture the color produced by one constituent may completely hide the others. For example, the wine-red color obtained from trinitrotoluene may prevent the identification of other nitrocompounds that may be present. The test may be made directly on a portion of the ether extract, since the presence of nitroglycerin does not in any way interfere. The sample under examination is dissolved in 2 to 3 c. c. of acetone or methyl alcohol, and a few drops of 10 per cent potassium or sodium hydroxide added. The characteristic colors produced by various nitrosubstitution compounds are shown in the table following.

---

a Gody, L., Traité théorique et pratique des matières explosives, 1907, p. 599.

Color reactions of nitrosubstitution products with alkalies.

Substance.	Form.	Color of solution.	Result of addition of alkaline solution.
Mononitrobenzene	Liquid	Colorless	No effect.
Dinitrobenzene	Crystal	Faint yellow	Purple-rose, turning to deep claret.
Trinitrobenzene	do	Colorless to pale yellow.	Rich purple-brown.
Mononitrotoluene (ortho)	Liquid	do	No effect.
Mononitrotoluene (meta)	Crystal	do	Slight yellow.
Mononitrotoluene (para)	do	do	No effect.
Dinitrotoluene	do	do	Gradual evolution of azure blue on standing.
Trinitrotoluene	do	do	Deep wine-red brown.
Mononitronaphthalene	do	do	No effect.
Dinitronaphthalene	do	Pale yellow	Reddish-yellow.
Trinitronaphthalene	do	do	Bright scarlet.
Tetranitronaphthalene	do	do	Reddish-yellow.
Picric acid	do	Golden-yellow	Precipitate of crystals of potassium picrate (orange).

Rosin, vaseline, paraffin, oils, etc., are found in the extract, after evaporation of the ether, as a dark-colored, greasy mass on the surface of the nitroglycerin or adhering to the walls of the beaker. Small amounts of resins and oils are generally found in the extracts from ordinary dynamite, these being normal constituents of the wood pulp, flour, or similar absorbents. The amount of such materials present is usually too small to be of any importance, and where only small traces are found they may best be disregarded. If it is desired to identify such materials the method for quantitative determination described later may be followed.

The water solution is examined for sodium, potassium, ammonium, zinc, the nitrate ion, etc. The qualitative and quantitative tests for these elements are discussed fully in all textbooks of analysis, and accordingly need not be repeated here. In ordinary qualitative work the writers have found the ring test for nitrates, with the use of the nitrometer in any doubtful case, to be most satisfactory; potassium and sodium are best identified by the color they impart to the flame of the Bunsen burner, viewed through at least two thicknesses of cobalt glass. The best test for ammonium is the evolution of ammonia by reaction with calcium oxide or sodium hydroxide. Small quantities of iron, aluminum, chlorides, sulphates, etc., are generally found in all dynamites as impurities from the raw materials used.

The residue insoluble in water is treated with cold dilute hydrochloric acid; effervescence, if noted, is an indication of the presence of a carbonate, and the usual tests are then made for CO_2. The acid extract after filtering is tested for calcium, magnesium, zinc, etc., whose carbonates or oxides may have been present in the powder as antacids.

The residue insoluble in water and dilute acid is best examined under the microscope, a magnification of 30 to 50 diameters being most convenient for all ordinary work. The identification of starch, cereal products, wood pulp, sawdust, kieselguhr, nitrocellulose, etc., is discussed in another part of this paper.

A starch test is readily made by heating a portion of the material to boiling with dilute acid, cooling and adding a drop of iodine solution, this operation producing an intense blue coloration if starch is present.

DETERMINATION OF MOISTURE.

The moisture present in dynamite may be determined in the following ways:

(a) By drying in a desiccator over sulphuric acid, calcium chloride, or other suitable drying agent without vacuum.

(b) By drying in a desiccator with vacuum.

(c) By passing dry air through the sample.

These methods have been carefully studied, and the first method, drying over sulphuric acid in a desiccator, has been found to have the widest application, and to be the most desirable in the analysis of explosives. In the determination of moisture in a sample of dynamite a number of important factors should be considered in order that accurate results may be obtained. The weight of the sample, the manner in which the sample is spread upon the watch glass, the size and type of desiccator, the exposed area of the desiccating agent, as well as its quantity and condition and the temperature at which the desiccation is carried out—all have an important bearing on the loss of weight resulting from desiccation, and unless care is taken moisture determinations made at different times on the same sample of dynamite will frequently give somewhat different results, owing to the slight variations in some of the above-named factors.

Preliminary studies were made to determine the amount of dynamite that is most suitable for the determination of moisture, and the thickness with which the sample should be spread upon the watch glass. The results of these tests showed that a sample of less than 2 grams of explosive is unsatisfactory, owing to the errors introduced, for example, by particles of material being carried away by the air during weighing, or upon opening the desiccator. When more than 2 grams of sample was taken the length of time required to effect thorough drying was unduly long; and consequently 2 grams of material, spread evenly on the concave surface of a 3-inch watch glass, was deemed the best amount for the determination. When only one or two analyses have to be run at a time, it is convenient to use a separate 5 or 6 inch desiccator for each sample, the bottom of the desiccator containing about 50 to 75 c. c. of concentrated sulphuric acid, and the watch glass being held by a triangle at a convenient and suitable distance above the acid. Objections have been raised [a] to the use of sulphuric acid as a desiccating agent in the determination of moisture in dynamite. The experience of the writers has shown these objections to be without grounds.

When a number of determinations are being made at the same time it is convenient to use larger desiccators, and the 10-inch size has been found to be satisfactory, five watch glasses being conveniently held at one time in such a desiccator.

IN ORDINARY DESICCATORS.

The period required for desiccation of a sample of dynamite in an ordinary desiccator is about 3 days or 72 hours. The first 12 hours within the desiccator reduces the moisture content to a very low percentage, but it has been found desirable to allow all samples to remain a further period of 60 hours, as the remainder of the moisture is slowly removed. The loss in weight, at the end of three days, of the 2-gram sample of dynamite on the watch glass is taken as moisture. After three days' drying, small amounts of moisture are still present, particularly in the wood pulp, but drying longer than three days has not been found desirable in ordinary analytical work. By long-continued desiccation, over either calcium chloride or sulphuric acid, a further gradual loss occurs which is evidently due to volatilization of nitroglycerin, and the amount of this loss varies with the temperature. Taking the loss of weight at the end of three days as representing moisture has been found to give uniform results in the case of dynamites of most varied composition, and has been adopted in the bureau's laboratory as a standard method, although it is impossible to say that at the end of any definite time all moisture has been removed from the sample and that further loss is due to volatilization of nitroglycerin.

Care should always be taken to spread the sample in a thin and uniform layer over nearly the entire surface of the watch glass, and the spreading should be done as quickly as possible to prevent undue exposure to the atmosphere. Experiments in which different sizes of desiccators (4-inch to 9-inch) were used showed that the rate of drying increased as the ratio of acid surface to powder surface increased, but that in three days the loss in all cases was practically the same. The effect of such variation is shown in the following table of results of tests of explosives containing amounts of moisture greater than that in ordinary dynamite. The table also shows a comparison between the efficiencies of two types of desiccators— the Scheibler, or ordinary type, which contains the acid in the bot-

---

[a] Gody, L., Traité théorique et pratique des matières explosives, 1907, p. 382; Guttmann, O., Schiess- und Sprengmittel, 1900, p. 162.

tom below the sample, and the Hempel type in which the acid is contained in the top or cover, above the sample. With an acid surface about 75 per cent as great as in the ordinary type the Hempel desiccator appears to be slightly more efficient than the Scheibler.

Results of moisture determinations showing effect of variation in acid surface and in type of desiccator.

Type of desiccator	Scheibler.	Scheibler.	Hempel.
Size of desiccator	4-inch.	9-inch.	6-inch.
Area of acid surface (approximate)	7 square inches.	50 square inches.	38 square inches.

Name of explosive.	Time of drying.	Moisture content determined by loss of weight.		
	Hours.	Per cent.	Per cent.	Per cent.
Explosive A [a]	6	5.73	9.84	9.97
	12	8.82	10.11	10.29
	24	10.06	10.33	10.43
	48	10.52	10.63	10.65
	72	10.66	10.68	10.79
Explosive B [a]	2	1.89	....	3.18
	5	4.09	....	4.88
	24	5.34	5.50	5.51
	48	5.49	....	5.60
	72	5.50	5.59	5.63

[a] Arbitrary designation of sample.

The quantity of acid used in the desiccator must also be considered, because the dilution of the acid by the absorbed moisture is dependent upon the amount of acid present. In general, the acid contained in a desiccator does not require frequent renewal, and when from 50 to 75 c. c. of concentrated sulphuric acid (H_2SO_4) is used in the desiccator the renewal of this material once each month or once every two months is all that is necessary.

The following experiment illustrates the effect of dilution of the sulphuric acid by the absorption of considerable moisture from powder samples. Two samples of the same dynamite were desiccated in similar desiccators, one containing 75 c. c. of 96 per cent sulphuric acid, the other the same volume of 90 per cent sulphuric acid. In three days the losses of moisture were 1.10 per cent and 1.02 per cent, respectively, and in five days 1.17 per cent and 1.05 per cent, respectively. In order that 75 c. c. of 96 per cent acid could become diluted to 90 per cent by the absorption of moisture from dynamite samples, it would be necessary that all of the moisture from about 460 2-gram samples of dynamite, each containing 1 per cent moisture, be absorbed by the acid. It is therefore obvious that the acid in the desiccators will not become diluted enough in one or two months to lose appreciably its affinity for moisture.

The effect of temperature on the moisture determination is shown in the following experiments, made with a 60 per cent dynamite. Three determinations of moisture were made on the same sample at widely different temperatures. Two grams of the sample in one desiccator were placed in an incubator, the temperature of which, by means of a thermostat, was regulated to a range of 36° to 38° C. (96.8° to 100.4° F.). A second desiccator containing another 2 grams of the sample was placed out of doors, the temperature varying during the experiment from $-23°$ to 2° C. ($-9.4°$ to 35.6° F.). A third desiccator containing a third 2-gram portion of the sample was left at room temperature, which varied from 17° to 25° C. (63° to 77° F.). In each of these experiments the 2-gram parts of the sample were spread upon 3-inch watch glasses, each watch glass being placed in a separate 4-inch desiccator containing 75 c. c. of fresh sulphuric acid. The results are tabulated below:

Results of tests to determine the moisture content of dynamite, showing the influence of temperature.

1	2			3	4	5		6	
Sample No.	Temperature.			Days exposed.	Loss of weight on desiccation.	Ether extract.		Nitroglycerin in ether extract.[a]	
	Maximum.	Minimum.	Mean.			By loss.	By direct weight.		
	° C.	° C.	° C.		Per cent.	Grams.	Grams.	Grams.	Per cent.
1	25	17	21	3 6 7 10 13	1.00 1.06 1.06 1.19 1.31	1.2158	1.2126	1.2051	60.26
2	38	36	37	3 6 7 10 13	1.09 2.49 2.74 3.54 4.12	1.1563	1.1522	1.1397	56.99
3	−7 −3 0 2 20	−16 −23 −14 −9 18	−9 19	3 6 7 10 13	0.60 .... 0.72 0.67 1.03	1.2243	1.2220	1.2114	60.57

a Determined by nitrometer.

After the weighings had been made on the thirteenth day the samples were carefully transferred to Gooch crucibles, and extracted with ether in a Wiley extractor.[a] The crucibles were then dried at 100° C., and the losses in weight determined, which are given in column 5. The nitroglycerin in each sample of ether extract was then determined by the nitrometer after the ether had been evaporated (see p. 35). The results shown under "loss of weight by desiccation" (column 4) indicate the pronounced influence which temperature has on the determination of moisture. In the case of sample 2 most of the moisture was probably lost within the first 24 hours, the further

---

a For the method used in the extraction see p. 30.

loss being entirely due to the evaporation of nitroglycerin. In the case of the sample 3 the weight first taken, at the end of three days, showed a loss of only 0.60 per cent. As previous experiments had shown that this amount of moisture can be removed from 60 per cent dynamite by about two hours' desiccation, it is probable that the loss noted during the first three days resulted during the first two hours' desiccation, before the sample became frozen. Practically no further loss occurred in this sample up to 10 days, and the loss which occurred between the tenth and thirteenth day was due to the fact that the sample had thawed. As the temperature in the laboratory frequently reaches 35° C. or more during the summer, it is evident that abnormally high results sometimes obtained during

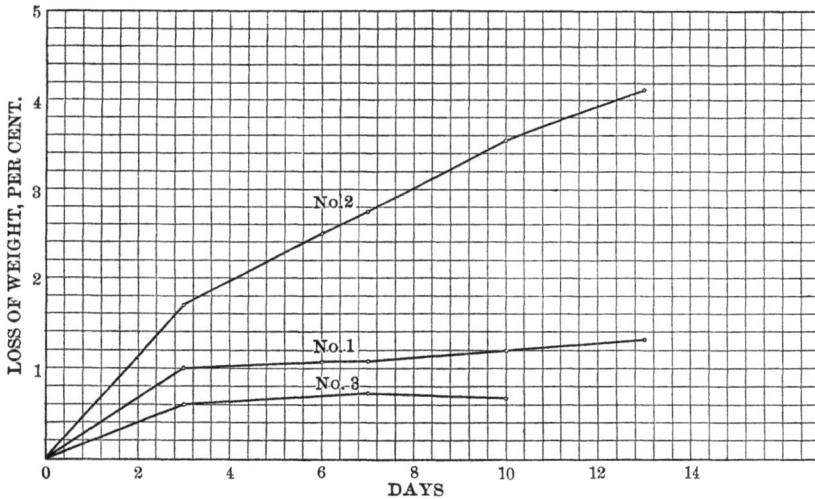

FIGURE 1.—Influence of temperature on determination of moisture in 60 per cent dynamite by desiccation over sulphuric acid. Temperature (mean): No. 1, 21° C.; No. 2, 37° C.; No. 3, —9° C.

warm days of summer may be in error to a considerable extent, being only in part due to loss of moisture and in part to volatilization of nitroglycerin. Since incorrect results are obtained when moisture is determined at a low temperature, no determinations of moisture should be made at a temperature sufficiently low to cause the sample to freeze. In order that perfectly uniform results may be obtained with samples of dynamite, the temperature of the room in which the desiccation is carried out should be practically constant at 20° C. This is a normal working temperature, and when moisture determinations are made at temperatures materially above or below it, allowance should be made for the influence of the abnormal temperature. The results given in the table on p. 22 are shown in the form of curves in figure 1. The curve for sample 3 represents the results of the first 10 days' exposure, during which period the sample remained frozen.

In the determination of moisture by the bureau's explosives laboratory, sulphuric acid has been accepted as the standard desiccating agent. As many chemists prefer to use calcium chloride, the following experiment is of interest, as showing the difference in efficiency of these two desiccating agents. Six 2-gram samples of 60 per cent dynamite were weighed from a large, well-mixed sample. Three of these samples were desiccated over calcium chloride and three over sulphuric acid. An individual 6-inch Scheibler desiccator was used for each sample. Weighings made at intervals showed the loss of moisture as follows:

Results of desiccation of 60 per cent dynamite over calcium chloride and over sulphuric acid, at room temperature.

Time.	Designation of sample.	Loss of weight over calcium chloride.	Designation of sample.	Loss of weight over sulphuric acid.
Days.		*Per cent.*		*Per cent.*
1/12	A	0.50	C	0.64
1/4	A	.59	C	.73
1/2	A	62	C	.79
3/4	B	.66	D	.87
1	A	.67	C	.84
2	B	.77	D	.94
3	A	.77	C	.96
3	B	.80	D	.97
3	E	.84	F	1.02
5	A	.84	C	1.01
5	B	.87	D	1.01
5	E	.91	F	1.07
6	A	.84	C	1.02
6	A	.87	D	1.05
6	E	.93	F	1.10
10	A	.95	C	1.11
10	B	1.00	D	1.17
10	E	1.00	F	1.20
13	A	1.07	C	1.23
13	B	1.10	D	1.32
13	E	1.15	F	1.36
16	A	1.16	C	1.33
16	B	1.13	D	1.46
16	E	1.25	F	1.48
19	E	1.25	F	1.53
20	E	1.37	F	1.60
34	E	1.72	F	2.01
50	E	2.22	F	2.55
75	E	2.82	F	3.35
111	E	3.67	F	4.55
183	E	5.77	F	7.55
237	E	6.85	F	9.20

These results, plotted in the form of curves for the two samples E and F, upon which desiccation was continued the longest, are shown in figure 2.

The fact that nitroglycerin is more or less volatile even at ordinary temperatures is recognized by most explosives chemists, and in some laboratories an endeavor is made to avoid loss of nitroglycerin during desiccation by keeping the atmosphere within the desiccator saturated with nitroglycerin vapors. This is done by placing in the desiccator, together with the sample of dynamite on which moisture is to be determined, a quantity of nitroglycerin on a watch glass, it being assumed that the evaporation of this nitroglycerin will saturate the atmosphere in the desiccator and minimize the loss of nitroglycerin from the sample of dynamite.

To test the value of such method the following experiments were made: Four samples (2 grams each) of the same 60 per cent dyna-

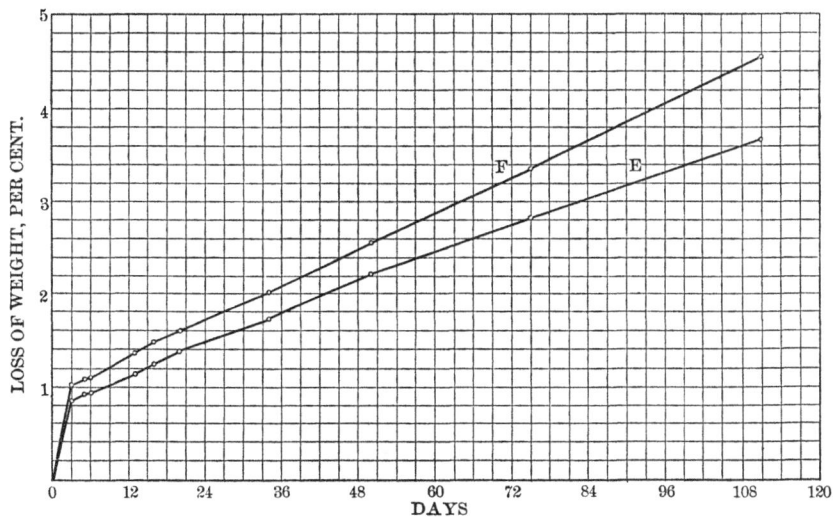

FIGURE 2.—Results of desiccation of 60 per cent dynamite over calcium chloride (E) and over sulphuric acid (F) at room temperature.

mite used above were spread on 3-inch watch glasses and placed in separate desiccators. Directly underneath each of the watch glasses was placed a watch glass containing a layer of fine dry sand saturated with nitroglycerin. Two of the desiccators contained sulphuric acid and two calcium chloride. Weighings of the dynamite samples at intervals showed losses of weight as follows:

Results of desiccation of 60 per cent dynamite over calcium chloride and over sulphuric acid in an atmosphere saturated with nitroglycerin vapors.

Designation of sample.	Desiccation over calcium chloride.		Designation of sample.	Desiccation over sulphuric acid.	
	Time.	Loss of weight.		Time.	Loss of weight.
	Days.	*Per cent.*		*Days.*	*Per cent.*
A	3	0.77	C	3	....
B	3	.80	D	3	0.92
A	6	.93	C	6	1.10
B	6	.97	D	6	1.12
A	10	1.02	C	10	1.13
B	10	1.00	D	10	1.15
A	13	1.15	C	13	1.20
B	13	1.17	D	13	1.25
A	19	1.25	C	19	1.31
B	19	1.35	D	19	1.40

Comparing these results with those of the table on page 24, it will be noted that the placing of an additional amount of nitroglycerin in the desiccator with the sample of dynamite has practically no effect on the amount of nitroglycerin lost from the sample.

The fact that nitroglycerin volatilizes in the desiccator in the presence of either sulphuric acid or calcium chloride suggested the carrying out of experiments to determine whether a sample of dynamite that had been desiccated for a sufficient time to lose all of its moisture would continue to lose weight in an empty desiccator (without any desiccating agent present).

Two 2-gram samples of 60 per cent dynamite were dried for three days on watch glasses over sulphuric acid, and then immediately

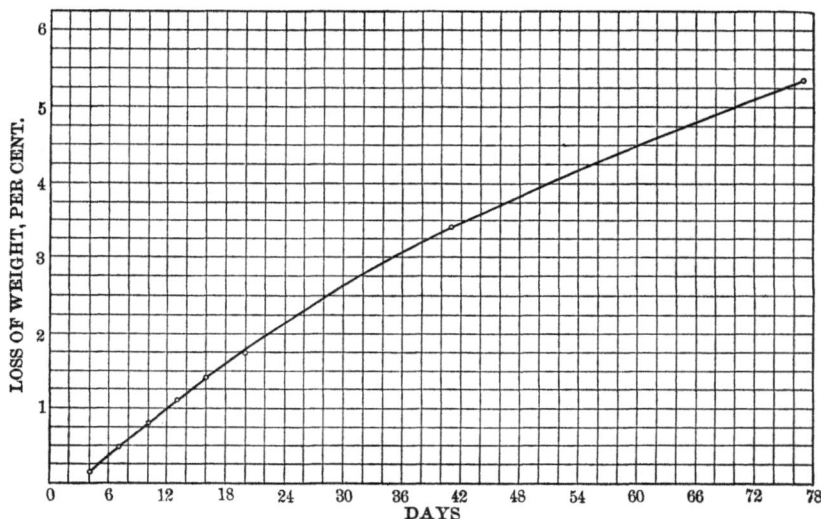

FIGURE 3.—Result of exposure of dry 60 per cent dynamite in a desiccator at 33° to 35° C. without desiccating agent.

placed in clean desiccators without any drying agent. One of these was kept at room temperature (17° to 22° C.), the other in a constant-temperature incubator oven at a temperature of 33° to 35° C. Weighings made at intervals showed a continued steady loss from the sample at 33° to 35°, whereas the sample exposed to room temperature gained in weight at first, then steadily lost until at the end of about 40 days it had attained its original weight. In the table below the percentage of loss or gain in weight recorded represents the variation from the dry weight of the samples after having been desiccated three days.

Change in weight of dried samples of 60 per cent dynamite in desiccators containing no desiccating agent.

Time in desiccator.	Change in weight.		
	Sample *A* (17°–22° C.).		Sample *B* (33°–35° C.).
	Gain.	Loss.	Loss.
Days.	*Per cent.*	*Per cent.*	*Per cent.*
4	0. 36	....	0. 13
7	. 35	....	. 46
10	. 30	....	. 80
13	. 25	....	1. 10
16	. 20	....	1. 40
20	. 20	....	1. 75
41	....	0. 05	3. 40
77	....	. 67	5. 35
202	....	2. 63	9. 33

It is often desirable to make a weighing at a shorter interval than 72 hours, and to obtain an approximation of the true moisture by calculation. An extended series of tests was made to determine what factor could be safely used, in conjunction with a weight taken at the end of 24 hours, to give the same result as would be obtained by desiccation for a period of 72 hours. It was found that the influence of the temperature of the room in which the desiccator was kept was so marked, and the difference in effect between different desiccating agents,[a] was such as to make this method uncertain. At the best only an approximation can be obtained, but when time is more important than accuracy a rough approximation to the true moisture content of ordinary dynamite can be obtained by considering the loss in weight over sulphuric acid in 24 hours to be 90 per cent of the total moisture. In other words, the loss of weight in 24 hours' desiccation multiplied by the factor 1.11 will be an approximation to the true moisture content. When large amounts of moisture are present, as in certain types of coal-mining explosives containing water added as a constituent of the explosive, it is not advisable to attempt the use of such a factor. In such cases the loss of weight on desiccating three days is considered as the total moisture.

IN VACUUM DESICCATORS.

The evaporation of water is much more rapid at reduced pressure than under atmospheric pressure, and therefore the determination of moisture may be made in a shorter time with the use of a vacuum than when atmospheric pressure prevails within the desiccator. If nitroglycerin were perfectly nonvolatile, desiccation in a vacuum would probably be sufficiently reliable for use as a standard method.

a See table on p. 24.

The following table shows the results obtained when samples of the same dynamite were exposed in desiccators of similar size and shape that contained the same desiccating agent, in one case a vacuum being used and in the other case the air within the desiccator being at the pressure of the air in the room.

Results of desiccating tests with and without vacuum in the desiccator.

Time dried.	Sample in ordinary desiccator.	Sample in vacuum desiccator.
	Per cent.	*Per cent.*
Loss of moisture in 24 hours	5.34	5.42
Loss of moisture in 48 hours	5.49	5.63
Loss of moisture in 72 hours	5.68	5.80

With the ordinary type of desiccator, containing acid in the bottom, moisture is removed from the sample much more rapidly with a vacuum than without, other conditions being equal. For example, in a 4-inch desiccator, with 7 square inches of acid surface on which the pressure was that of the atmosphere, the loss of moisture from a powder sample containing about 11 per cent of water was only 54 per cent of the total in 6 hours, 83 per cent in 12 hours, and 94 per cent in 24 hours. In a slightly larger desiccator of the same type, with 12 square inches of acid surface, above which there was a vacuum, the loss in 6 hours was 91 per cent, in 12 hours 95 per cent, and in 24 hours 96.5 per cent. In both samples a practically constant value was obtained in 3 days.

In the case of the Hempel desiccator, containing acid above the explosive sample, the rate of loss of moisture is nearly the same with or without vacuum.

In general it may be stated that any type of vacuum desiccator will remove in 12 hours practically all the moisture from dynamites and even from permissible explosives containing as much as 10 to 12 per cent of water, the result thus obtained being about the same as that given by 3 days' desiccation without vacuum.

Vacuum desiccation for a longer time than 12 hours will cause a further slight loss, which is probably for the most part due to volatilization of nitroglycerin.

IN A DRY-AIR CURRENT.

When a current of air previously dried by being passed through a calcium chloride tube, or through sulphuric acid, is passed through a mass of dynamite in a suitable sample tube, the dynamite gives up its moisture to the dry air. Drying may be very quickly effected in this manner, because of the fact that the dry air comes into intimate contact with all portions of the explosive, but the method has the

serious defect of being inaccurate, owing to the volatility of nitro-glycerin.

In desiccation by means of dry air, a drying tube of the form shown in figure 4 is usually employed. Sometimes several of these tubes are connected together in a train. The dry air is always intro-duced at the bottom, so that it may rise through the explosive, and a weighed sample of about 15 grams is generally taken for the deter-mination.

Several tests were made to determine the accuracy of this method of drying samples of dynamite. A current of air dried by passing through sulphuric acid was passed through a sample of dynamite which had been found, by desiccation for three days over sulphuric acid, to contain 5.68 per cent of moisture. Weighings at intervals showed loss of weight as follows.

Loss of weight of dynamite in a current of dry air.

Time dried.	Loss of weight.
Hours.	*Per cent.*
2	1.47
6	4.29
12	5.20
24	5.50
48	5.70
60	5.80
72	5.86

The table shows that at the end of 48 hours there had been a loss in weight approximately corresponding to the percentage of moisture found in the explosive by the standard method. The constant loss after that point represents volatilization of nitroglycerin, although it is of course to be noted that the loss of nitroglycerin was continu-ous, nitroglycerin being removed from the time the air first began to pass through the sample. How inaccurate this method is may be readily understood when one remembers that nitroglycerin can be completely volatilized by bubbling dry air through it for a sufficient length of time, and that accordingly any sample of dynamite would probably reach an ultimate value in which the loss in weight would correspond to the amount of moisture plus the amount of nitroglyc-erin present in the sample.

SUMMARY.

The most satisfactory method for the determination of moisture consists in using a 3-inch watch glass containing an evenly distrib-uted 2-gram sample, the watch glass and sample being placed in a desiccator containing sulphuric acid, and being weighed at the expira-tion of 72 hours. Care should be taken that the temperature does not vary widely from a mean of 20° C. An approximation of the true

moisture content of the sample of explosive may be obtained by desiccating in the manner just described for a period of 24 hours, and
multiplying the loss in weight thus found by the factor 1.111, or the
loss in a vacuum desiccator for 24 hours may be taken without further calculation as an approximation to the true moisture content
as determined by standard methods.

EXTRACTION WITH ETHER.

Extraction with ether removes from dynamite not only nitroglycerin, but also any resins or sulphur that may be present. Aside
from resin intentionally admixed
there is always some resin present
in the wood pulp. In addition to
these constituents, in dynamite
small amounts of oil are sometimes
found, having been introduced from
mixing machines or packing machines. When flour, corn meal, or
other grain or cereal products are
present as constituents of explosives a small amount of oil from
such materials is also found in the
ether extract. Nitrotoluenes, paraffin, vaseline, etc., are not normal
constituents of ordinary dynamite,

FIGURE 4.—Drying tube.

and the determination of these substances is discussed under the
type of explosive in which they occur as characteristic components,
but it is to be noted that were any of these materials present in
dynamite they would be found in the ether extract.

REFLUX-CONDENSER METHOD.

A sample of from 6 to 10 grams of the explosive is weighed in either
a porcelain Gooch crucible with asbestos mat or a porous alundum [a]
crucible of about 25 c. c. capacity. When a Gooch crucible is used
the mat should be light, but should be perfectly coherent. Such a mat
is prepared in the bureau's explosives laboratory as follows: Five
grams of short-fiber asbestos, in the form of short shreds, free from
hard lumps, is added to 1 liter of water. When used the mixture is
well shaken and about 10 c. c., an amount sufficient to fill the crucible
to about two-thirds of its depth, is poured into the crucible. Suction is applied, and a smooth and perfect mat is almost invariably
produced. The crucible and mat are then carefully dried for some
hours at 100°, weighed, and placed in a desiccator. For extracting
with ether, the form of apparatus found most satisfactory is the

[a] A trade name for artificially prepared porous aluminum oxide.

Wiley extractor, as shown in Plate II, *A*. The crucible is supported on a small hanger made by twisting No. 18 copper wire into suitable shape.

In performing an extraction the crucible, with its weighed sample of explosive, is placed in the hanger, and about 35 c. c. of U. S. P. ether (96 per cent) is poured in several portions through the sample into the glass extraction tube. Water is continuously circulated through the cooling coils of the condenser. The ether is boiled by means of an electric heater or a vessel of hot water in which the lower part of the tube is immersed. The ether vapor condenses on the surface of the metal condenser, the condensed ether dropping into the crucible and percolating through the sample of explosive. The temperature is regulated so that the sample will be kept covered with ether without any overflow.

When a vessel of hot water is used for heating the tube, the ether partly drains out of the crucible during a change of water, but is at once replaced by a fresh supply. This intermittent action probably accomplishes a more efficient extraction than is obtained by keeping the sample continuously covered with ether. The extraction with ether is continued for about three-fourths of an hour for most explosives. This period is usually somewhat longer than is necessary to remove all of the nitroglycerin, but it is desirable to carry on the extraction long enough to insure the complete removal of material soluble in ether, so as to avoid testing for completeness of extraction. If for any reason a shorter time for extraction is desirable, the extraction is continued for the time desired, after which a small additional amount of ether is put into the apparatus, and the ether passing through the crucible is evaporated in a watch glass and an examination made for residue. If residue is found, complete extraction has obviously not been accomplished.

The crucible containing the portion of the explosive insoluble in ether is placed in a drying oven heated to about 100° C. This should be done promptly, since the evaporation of the ether with which the contents of the crucible are saturated lowers the temperature of the crucible sufficiently to cause the precipitation of considerable moisture upon the crucible and its contents, and such a precipitation is undesirable as it necessitates longer drying. Although no loss or inaccuracy in analysis is liable to result from the constituents of the explosives becoming wet at this stage, yet for uniform results in drying it is usually best to transfer the crucible directly from the Wiley extractor to the drying oven. To avoid filling the drying oven with ether vapors, it is convenient to have a suction flask and carbon tube near the Wiley extractor, and as soon as the ether extraction is completed the crucible, still very wet with ether, may be placed in the carbon tube and sucked dry, after which it is placed in the drying oven.

If the qualitative examination has indicated the presence of ammonium nitrate, the drying of the material insoluble in ether should be carried out at 70° C. instead of 100° C., because at 100° ɩn appreciable loss of ammonium nitrate results whereas at 70° the loss is slight.

The periods of drying generally adopted are five hours at 95° to 100°, and overnight or 18 to 24 hours at 70° C. Even at the higher temperature no error results by drying overnight unless ammonium salts or other volatile ingredients are present. A shorter time than five hours is probably sufficient in most cases, but the five-hour period has been adopted to cover all cases and obviates any necessity of an additional check weighing.

The loss of weight represents all ether-soluble material plus the moisture originally present in the sample.

The ether extract is transferred from the glass extraction tube to an evaporating dish of low pattern or to a small beaker previously weighed. The extraction tube is then washed out with a small quantity of pure ether, which is added to the ether extract in the evaporating dish. The contents of the evaporating dish are allowed to evaporate spontaneously; a number of hours are usually allowed for the evaporation, the best results being obtained when the period is overnight. After the ether has evaporated, the residue is thoroughly dried by leaving the dish for a few hours in a vacuum desiccator. The weight is then noted; it is usually a little less than the total loss on ether extraction minus the moisture as determined by desiccation. The difference is due to volatilization of the nitroglycerin during the evaporation of the ether, and is considered later.

A more nearly correct value for the weight of material removed by ether is obtained by deducting the amount of moisture determined by desiccation from the total loss of weight found by extraction, the direct weight of the ether extract after the evaporation of ether being used only as a check. In all cases ether extraction should be made in duplicate, one sample of the weighed extract being used for the determination of the nitroglycerin with the nitrometer, and the other sample being used in determining the other constituents present.

SUCTION METHOD.

In the laboratories of some dynamite works extraction with ether is made without any form of continuous-extraction apparatus; the sample in the Gooch crucible is merely washed several times by pouring ether through it, applying suction after each addition of ether. This method involves the use of greater quantities of ether,

A. APPARATUS FOR ETHER EXTRACTION. WILEY EXTRACTOR.

B. GRAVIMETRIC BALANCE.

and the objection has been made that the reduction of temperature resulting from the evaporation of ether causes a deposition of moisture from the air current drawn through the sample, this moisture dissolving out small amounts of the water-soluble nitrates that pass through with the next addition of ether.

To test the merits of this method in comparison with the usual extraction method, ether extractions were made on a large number of samples of 45 per cent dynamite, using both the reflux-condenser method and the suction method. In the latter method about 100 c. c. of ether in six portions was passed through each sample, each portion of ether being allowed to stand in the crucible for one minute before suction was applied. The suction was continued for periods of one-half minute to two minutes in order that different amounts of air might be drawn through the samples. The samples were then dried and weighed as usual.

In general, extraction in the Wiley apparatus gave slightly lower results than the suction method, although in most cases the difference between duplicate samples extracted by the same method was as great as the variation between the two methods. This fact is explained by the lack of homogeneity of the dynamite. For example, a few unusually large particles of wood pulp or nitrate in one sample may cause a greater variation in the percentage of ether extract than the variation actually due to the method of extraction employed.

COMPARATIVE EXTRACTIONS WITH ANHYDROUS AND U. S. P. (96 PER CENT) ETHER.

To ascertain the effect of the purity of the ether used for extractions, determinations of ether extracts were made on several types of explosives, and on various carbonaceous absorbents used in dynamites. Duplicate determinations were made using both anhydrous ether (distilled over sodium) and U. S. P. (96 per cent) ether.

Results of extractions are shown in the following table, the values given being the percentage of loss of weight noted on weighing the insoluble portion after drying five hours at 95° to 100°. The loss of weight in each case therefore includes any moisture originally present.

Loss of weight of explosives and of carbonaceous absorbents by ether extraction.

	Kind of ether.	
Kind of sample.	U. S. P. (96 per cent).	Anhydrous.
	Per cent.	*Per cent.*
Dynamite (A)	25.86	25.69
Dynamite (B)	32.58	32.51
Dynamite (C)	40.77	40.74
Corn meal	14.81	14.92
Wheat flour	13.18	13.12
Wheat middlings	14.91	14.78
Corn meal, dry	1.81	1.71
Wood pulp, dry (1)	2.71	2.18
Wood pulp, dry (2)	2.59	2.20
Wood pulp, dry (3)	2.31	1.85
Wood pulp (4)	8.84	8.02
Wood pulp (5)	5.44	4.73
Wood pulp (6)	5.98	5.48
Wood pulp (7)	12.24	11.98
Wood pulp (8)	13.30	12.75
Wood pulp (9)	5.65	5.23
Wood pulp (10)	6.23	5.88
Wood pulp (11)	5.01	4.85
Wood pulp (12)	7.74	7.40
Wood pulp (13)	8.27	7.86

From the table it is apparent that U. S. P. ether extracts a larger percentage of material from the commonly used carbonaceous absorbents than does anhydrous ether, which is practically free from alcohol. It is probably because of the alcohol present, in amounts up to about 4 per cent, that U. S. P. ether shows the greater extractive power.

EFFECT OF MOISTURE IN DYNAMITE ON EXTRACTION WITH ETHER.

Numerous authorities prescribe that the ether extraction shall be made on a sample previously dried to constant weight in a desiccator.[a] Presumably this specification is aimed to prevent water-soluble constituents from being carried through in the ether extract. Such an error is naturally greater as the amount of moisture present in the dynamite is greater. Accordingly experiments were made on mining explosives similar to ordinary dynamite, to which water had been added as an additional constituent for the purpose of reducing the temperature of explosion. Two explosives containing 10.70 and 5.70 per cent of moisture, respectively, were extracted with ether in the usual manner, (1) in the original condition, and (2) after having been dried in vacuum desiccators over sulphuric acid for 24 hours. The amounts of materials soluble in ether extracted are shown in the following table.

---

[a] Guttman, O., Schiess-und Sprengmittel, 1910, p. 162; Kedesdy, E., Sprengstoffe, 1909, p. 247; Escales, R., Die Explosivstoffe, vol. 3, 1908, p. 204.

Results of ether extraction of two explosives in different conditions.

1	2	3	4	5
Sample.	Condition.	Moisture.	Extract.	Difference.
		Per cent.	*Per cent.*	*Per cent.*
A	1	10.70	26.52	0.59
	2	.44	a 25.93	
B	1	5.70	25.79	.36
	2	.45	a 25.43	

a Calculated to explosive in original condition.

The amount of moisture present in the dried samples was found by desiccating portions of each sample for a further period of three days in ordinary sulphuric-acid desiccators.

The values in column 4 represent the total loss on extraction less the moisture content of the sample, all results being expressed as a percentage of the amount of original undried explosive.

Assuming that the extracts from the samples in their original condition (condition 1) are larger than those from the dried samples (condition 2) because of loss of water-soluble nitrate in the moisture taken up by the ether, it is apparent that in the case of ordinary dynamite containing only one or two per cent moisture any loss from this source is negligible.

DETERMINATION OF NITROGLYCERIN.

The nitrogen of organic or inorganic nitrates or nitrites is readily evolved as nitric oxide (NO) by reaction with sulphuric acid and mercury in the nitrometer. A determination of such nitrogen in the extract therefore serves as a means of calculating the amount of nitroglycerin present. The form of nitrometer found by the authors to be most satisfactory for explosives work is the modified Lunge nitrometer, as illustrated in Plate III.

THE NITROMETER.

This instrument a consists of six glass parts as follows: A globe-shaped reservoir (a); a generating bulb (b) of about 300 c. c. capacity, the generating bulb having stopcocks at both top and bottom to permit a violent agitation, and having a cup above which communicates with the bulb through the upper stopcock; a second globe-shaped reservoir (c), to which, by means of a glass multiple connecting tube and rubber tubing, are joined a compensating burette (d), a reading burette (e), and an additional measuring burette (f). The reading and compensating burettes are of the same shape and size, and

a The description has been taken in a large part from a paper by J. R. Pitman on The analysis of nitric and mixed acids by du Pont's modification of the Lunge nitrometer, Jour. Soc. Chem. Ind., vol. 19, 1900, p. 983; see also Lunge, G., Du Pont's nitrometer, Jour. Soc. Chem. Ind., vol. 20, 1901, p. 100.

are blown out into bulbs at the top. The compensating burette is not graduated. Above the bulb it has a small vertical open tube, which is sealed when the instrument is standardized. The reading burette is calibrated so that percentages of nitrogen may be read therefrom, and is marked to read from 10 to 14 per cent, being graduated to one-hundredths of 1 per cent. Between 171.8 and 240.4 c. c. of gas must be generated to obtain a reading; that is, the 10 per cent mark represents the volume of 171.8 c. c. of NO at 20° and 760 mm. pressure, containing 0.1 gram of nitrogen; the 14 per cent mark is equal to 240 c. c. NO under the same conditions, representing 0.14 gram nitrogen.

The compensating burette is supported by a ring; the generating bulb is supported just above each stopcock by forked holders, curved so as to retain the bulb in place. In order to remove the generating bulb it needs only to be raised slightly and brought forward, the manipulation of a screw, as with an ordinary clamp, being thus avoided. The two reservoirs and the reading burette are supported by ring clamps, these clamps having milled rollers at the shank; they are moved up and down vertical racks by means of hand screws, the rollers being so arranged in conjunction with the vertical racks that the weight of the part presses them down and acts as a brake, thus preventing their moving when not being manipulated.

Having arranged the apparatus and filled the compensating, reading, and generating tubes as well as their connections with mercury, the next step is to standardize the instrument. Twenty to thirty cubic centimeters of sulphuric acid is run into the generating bulb through the cup at the top, and at the same time about 210 c. c. of air is let in; the cocks are then closed and the bulb is well shaken; this shaking thoroughly desiccates the air, which is then run into the compensating burette until the murcury is about on a level with the 12.50 per cent mark on the reading burette, the two burettes being held at the same height. The compensating burette is then sealed off at the top. A further quantity of air is desiccated in the same manner and run over into the reading burette until the height of mercury in the reading burette stands at about the 12.50 per cent mark. The cocks are then closed, and a small piece of glass tubing, filled with sulphuric acid (not water), and bent in the form of a U, is attached to the outlet of the reading burette. When the mercury columns are about balanced and the inclosed air has been cooled to room temperature, the cock is again carefully opened, and when the sulphuric acid balances in the U tube, and the mercury columns in both burettes are therefore at the same level, the air in each tube is subject to the same conditions, namely, atmospheric temperature and pressure. A reading is now made from the burette,

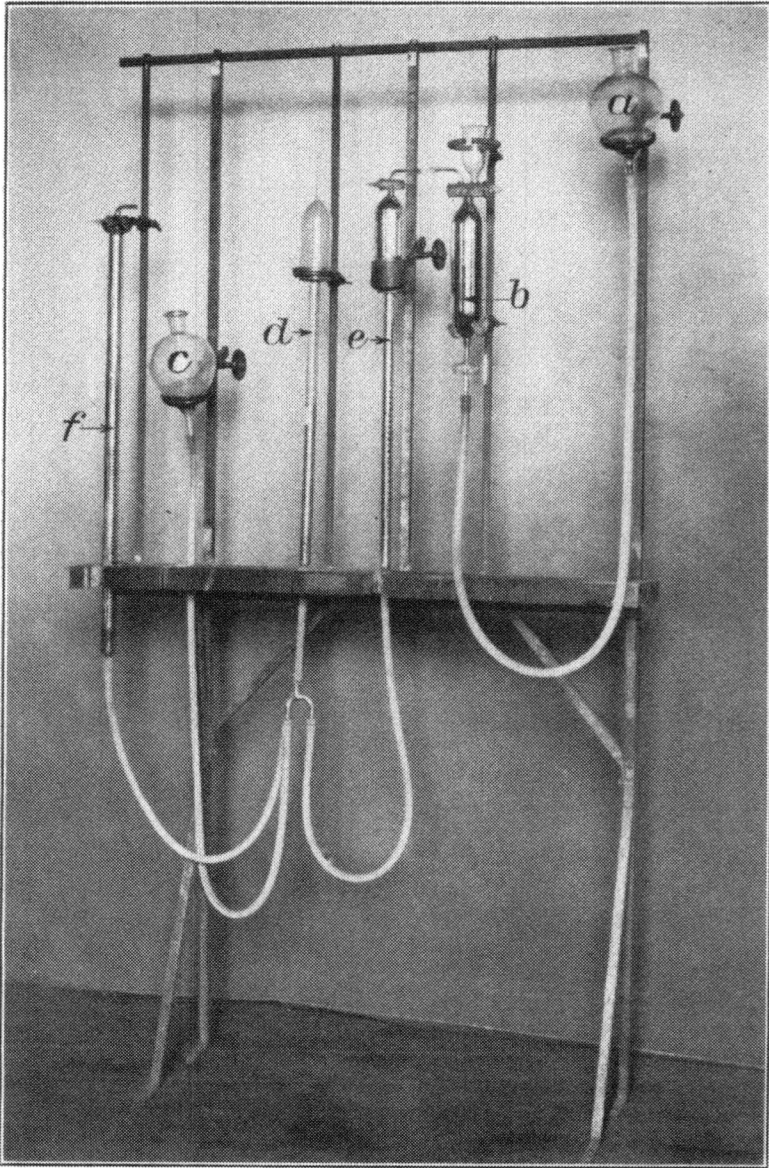

NITROMETER.

and the barometric pressure and temperature are carefully noted. Using the well-known formula

$$V = \frac{V'P'(273+t)(1-.00018t')}{P(273+t')(1-.00018t)}$$

the volume this inclosed air would occupy at a pressure (P) of 29.92 inches of mercury (760 mm.) and at a temperature (t) of 20° C. is determined. The cock is again closed and the reservoir and reading burette carefully adjusted so as to bring the air in the reading burette to the calculated volume and the mercury in the compensating burette to the same level as the mercury in the reading burette. A strip of paper is now pasted on the compensating burette at the level of the mercury, and the standardization is then complete.

There is, however, another and shorter method of standardization than the one described above. It is well known that the quality of the sulphuric acid used in the nitrometer will materially affect the results. To ascertain whether sulphuric acid is suitable for use in making nitrogen determinations in the nitrometer a determination is made on chemically pure dry potassium nitrate and the reading obtained in the nitrometer is compared with the theoretical percentage of nitrogen in potassium nitrate. In applying this procedure to the standardization of the nitrometer the compensating burette is filled with desiccated air, as described above, and 1 gram of potassium nitrate, dissolved in 2 to 4 c. c. of water, is introduced into the generating bulb, the cup is washed with 20 c. c. of 95 to 96 per cent sulphuric acid in three or four portions, and each portion is run separately into the bulb. The gas, when generated, is run over into the reading burette, and the mercury columns in both burettes are leveled, so that the mercury in the reading burette is also at 13.87, the theoretical percentage of nitrogen in potassium nitrate. A strip of paper is pasted on the compensating burette at the level of the mercury, and the standardization is then accomplished.

This method of standardizing offers many advantages over that first described, among which may be mentioned that no readings of temperature or pressure are necessary. Probably the greatest advantage is that if the acid used in standardizing should contain impurities, which might otherwise affect the result, the error is entirely compensated and corrected in subsequent work; that is to say, the instrument having been so standardized that the reading gives the theoretical percentage of nitrogen in potassium nitrate, the results will be accurate when testing other substances so long as the same quantity of sulphuric acid from the same lot is used.

It must, of course, be understood that once having standardized the instrument with a certain lot of acid no different lot of acid can be used without restandardizing. In order to avoid slight differences in results due to variations in the acid, it is advisable to reserve a sufficiently large uniform stock of acid, for example, a carboy full, for nitrometer use.

The additional measuring burette, with which this type of nitrometer is provided, known as the "universal tube" (f, Pl. III), is simply a straight burette, marked to read from 0 to 100 in percentages and graduated to one-tenth of 1 per cent. The tube is of such a size that 0.30 gram of NO (or $\frac{1}{100}$ gram-molecule of NO) under standard conditions of temperature and pressure (20° and 760 mm.) fills it to the 100 mark.

If it is desired to read the percentage of nitrogen direct, 0.14 gram of substance is weighed out; if the percentage of NO_2 is desired, 0.46 gram of substance is weighed out. Consequently, if 1.01 grams of potassium nitrate, 0.63 gram of nitric acid, or 0.85 gram of sodium nitrate are used, the results can be read directly as percentages of the original substance.

This method is convenient when it is not certain that the reading will fall within the limits of the graduations in the ordinary measuring burette.

The "universal tube" is found particularly advantageous when, for example, the amount of nitroglycerin in a sample is so small that the volume of gas generated is insufficient to fill the large reading burette to its graduated portion. The volume of gas generated from any amount of nitroglycerin up to about 0.75 gram may be read in the "universal tube." Readings in this measuring tube can be as accurately made as in the regular reading burette.

PROCEDURE.

To determine the amount of nitroglycerin in the ether extract of a dynamite, the sample from which the ether has been evaporated is dissolved in 5 to 10 c. c. of sulphuric acid (specific gravity, 1.84) and transferred to the generating bulb of the nitrometer, the beaker and the cup of the nitrometer being rinsed with several further additions of acid until 20 to 25 c. c. has been used. If the quantity of nitroglycerin present is too great, the sample dissolved in sulphuric acid is transferred to a burette and an aliquot part run into the nitrometer cup and washed into the generator with about 20 to 25 c. c. of sulphuric acid. The maximum amount of pure nitroglycerin used should be not greater than 0.75 gram in order that the gas generated will not exceed the volume of the reading burette.

The generator is then shaken gently until the generation of gas has forced out all but about 60 to 75 c. c. of the mercury, the reservoir being lowered if necessary in order to reduce the amount of mercury to this extent. The cock at the bottom of the generator is then closed and the generator shaken violently for about two to three minutes. After allowing all bubbles to separate from the reaction mixture, the gas is transferred to the reading burette, the surface of the mercury in the burette is brought to the same level as that in the compensating burette when the dry air in the compensating burette occupies the standard volume indicated by the strip of paper attached in calibrating.

The gas is allowed to stand for a few minutes to obtain an equilibrium of temperature, the levels being readjusted if necessary, and the reading is noted. This reading divided by 18.50 equals the weight of nitroglycerin in the sample used for the determination.

A more or less serious error to be considered in the determination of nitroglycerin is that introduced by losses due to volatilization of the nitroglycerin during the evaporation of the ether. To determine the effect of the rapidity of evaporation on the amount of nitroglycerin lost, weighed samples (0.6 to 0.7 gram) of nitroglycerin were placed in 100 c. c. beakers, treated with 50 c. c. of ether, the ether evaporated at different rates, and the samples dried in vacuum desiccators to remove the moisture taken up during the evaporation of the ether. Nitrogen was then determined by means of the nitrometer, the weight of nitroglycerin being calculated from the nitrometer reading. The results obtained are tabulated below:

Loss of nitroglycerin on evaporating ether extract.

[Determinations by J. H. Hunter.]

1	2	3	4	5	6	7
Sample.	Original weight of sample.	Weight of residue after evaporation of ether.	Nitrometer reading.	Weight of nitroglycerin found by nitrometer.[a]	Loss of nitroglycerin (2-5).	Method of evaporation.
	Gram.	Gram.	Per cent.	Gram.	Gram.	
1.........	0.6469	0.6458	11.73	0.6389	0.0080	At room temperature overnight.
2.........	.6328	.6318	11.53	.6280	.0048	Do.
3.........	.6212	.6140	11.17	.6084	.0128	Gentle boiling.
4.........	.6643	.6375	....	....	b .0268	Do.
5.........	.6199	.6167	11.31	.6160	.0039	Do.
6.........	.6666	.6660	11.96	.6514	.0152	Current of compressed air blown over beaker.
7.........	.6773	.6710	12.10	.6590	.0183	Do.

a Weight of nitroglycerin=nitrometer reading÷18.50.   b Loss of weight.

No attempt was made to obtain constant weight after evaporation of the ether, the samples being left in vacuum desiccators only long enough to remove most of the water; hence the weights in column 3 are greater than the weights of nitroglycerin calculated from the nitrogen found (column 5).

The figures in column 6 represent the differences between the weights in columns 2 and 5.

It was noted that rapid removal of the ether, either by means of gentle heating or by means of an air current, caused a greater loss of nitroglycerin than did slow spontaneous evaporation at room temperature, the only exception being in the case of sample 5, with which the loss of nitroglycerin was only 0.0039 gram, the ether being volatilized by gentle boiling. The large loss noted with sample 4 was probably due to spurting.

Such losses as are shown in the table do not greatly affect the determination of nitroglycerin in a sample of dynamite. Thus, in analyzing a 6-gram sample of dynamite, a loss of 0.01 gram of nitroglycerin would be equivalent to only 0.17 per cent of the original sample. The importance of the error is of course greater as the percentage of nitroglycerin in the sample is less.

Evaporation in the bell-jar evaporator.—An improved method of removing the ether from the ether extract without appreciable loss of nitroglycerin was devised by A. L. Hyde in the bureau's laboratory. The beaker containing the ether solution is placed on a ground-glass plate and covered by a glass bell jar about 6 inches in diameter and 8 inches high, having two tubulures, one at the top and one on the side, each opening being fitted with a perforated stopper and delivery tube. A rapid current of compressed air, dried by passage through concentrated sulphuric acid, in two wash cylinders, is allowed to enter through the glass tube in the top of the bell jar, the lower end of the tube being about one-half inch above the surface of the ether solution in the beaker. The air current is so regulated that a marked "dimple" is made in the surface of the solution, care being taken to prevent any loss by spattering. The possibility of acid being mechanically carried over from the cylinders is avoided by connecting an empty trap between the cylinders and the bell jar. The ether vapors pass out through the glass tube in the side tubulure and may be conducted out of the laboratory through a rubber tube passing to a hood or out of a window.

The low temperature produced by the rapid evaporation of the ether minimizes the volatilization of the nitroglycerin, and the fact that the air is thoroughly dried prevents any deposition of moisture in the beaker, so that it is not necessary to desiccate the residue after the ether has entirely volatilized.

The following preliminary tests show the efficiency of the method: A weighed quantity of nitroglycerin was dissolved in 50 c. c. of ether, the ether evaporated as described, and the residue in the beaker weighed at intervals.

Loss of nitroglycerin by evaporation of ether solution in bell-jar evaporator.

Time of evapora-tion.	Weight of sample.			
	A	B	C	D
Hours.	Grams.	Grams.	Grams.	Grams.
0	a 2.838	a 3.236	a 2.979	a 2.620
2	3.176	3.359	3.189	2.801
3	2.941	3.286	....	....
4	2.881	3.262	3.037	2.644
5	2.863	3.251	....	....
6	2.856	3.245	2.987	2.623
7	2.851	3.241	2.984	2.621
9	2.843	3.236	2.978	2.619
11	2.841	3.235	2.977	2.618
14	2.837	3.234	....	....

a Original weight of nitroglycerin.

U. S. P. ether (96 per cent) was used in tests A and B, and alcohol-free ether (distilled over sodium) in tests C and D. When the 96 per cent ether was used a distinct odor of acetic aldehyde was noticed and the rate of loss of weight was slightly lower, due no doubt to the presence of alcohol in the ether.

It is apparent from the above results that this method offers a convenient and rapid method of removing the ether without appreciable loss of nitroglycerin. Evaporation for about six hours removes the ether sufficiently to permit determination of the nitroglycerin in the nitrometer.

DETERMINATIONS OF SULPHUR, RESINS, ETC.

The sulphur used in dynamite is the form known as crushed brimstone. It is soluble in about 100 parts of ether at 23.5° C.,a and unless present in large amount in the sample of explosive being analyzed it will all be removed by the extraction with ether. However, when a considerable amount of sulphur crystallizes out in the ether extract, it is always advisable, after the water extraction, to make a further extraction of the explosive with carbon disulphide, in order to insure the complete removal of the sulphur.

As already mentioned, the analysis of an explosive is carried out in duplicate, one sample of the ether extract being used for the determination of nitroglycerin, the duplicate sample being used for the determination of sulphur, resins, etc. The duplicate sample is treated as follows: The weighed extract is redissolved in a mixture of ether and alcohol previously neutralized with standard alkali. The solution thus obtained is titrated with standard alcoholic potash to determine resins, phenolphthalein being used as an indicator. Determinations of a number of samples of commercial rosin (colophony) gave rather uniform results, 1 c. c. of normal alkali being found

a Gody, L., Traite théorique et pratique des matières explosives, 1907, p. 85.

equal to 0.34 gram of rosin, which agrees with the value given by Lewkowitsch.[a]

After titration a large excess of alcoholic potash is added and the mixture is heated on the steam bath, preferably overnight, in order to saponify the nitroglycerin. It must be remembered that nitroglycerin so treated saponifies slowly. Hence the reaction must not be hastened by heating to a higher degree than that obtained on a water bath, as an explosion may result. When saponification is complete the residue left upon evaporation is shaken with water and ether and separated in a separatory funnel. Any oily material (vaseline, paraffins, etc.) that can not be saponified is dissolved in the ether and may be weighed after evaporation. The water solution is acidified with hydrochloric acid and treated with bromine to oxidize the sulphur. Any rosin that was originally present will have formed a soap with the alkali; the acid decomposes this soap, and the rosin separates out from the acid liquid, floats on it, and may be readily removed, dried, and weighed, the weight serving as a check on the results of titration. The sulphur is oxidized to sulphuric acid by the bromine and may be determined by precipitation as barium sulphate.

Sulphur may be separated from nitroglycerin by a method depending on the fact that nitroglycerin is soluble in 70 per cent acetic acid, whereas sulphur dissolves only slightly in either glacial or 70 per cent acetic acid. The extent to which sulphur dissolves in acetic acid was determined by experiments with both brimstone and flowers of sulphur, in both cases the material being pulverized so as to pass through an 80-mesh sieve.

One gram of sulphur was digested in 100 c.c. of acetic acid for a definite period of time; the mixture was then washed on to a weighed Gooch crucible, dried for five hours at 70°, and weighed. The loss in weight represented the amount of sulphur dissolved by 100 c. c. of acid. The results were as follows:

Solubility of sulphur in acetic acid.

[Determinations by J. H. Hunter.]

Form of sulphur.	100 c. c. of 70 per cent acetic acid.			100 c. c. of glacial acetic acid.		
	Temperature.	Time.	Weight of sulphur dissolved.	Temperature.	Time.	Weight of sulphur dissolved.
	° C.	Hours.	Gram.	° C.	Hours.	Gram.
Brimstone..........	25	1	0.000	25	1	0.0338
	25	25	.000	25	1	.0354
	80	1	.0226	80	1	.1658
	80	1	.0127	80	1	.1464
Flowers of sulphur..............	25	20	.0115	....	....	....
	25	20	.0090	....	....	....
	80	1	.0027	....	....	....
	80	1	.0036	....	....	....

a Lewkowitsch, J., Chemical technology and analysis of oils, fats, and waxes, vol. 1, 1909, p. 502.

These determinations show that sulphur (brimstone) can be separated from nitroglycerin by means of acetic acid (70 per cent) at ordinary room temperature without appreciable loss of the sulphur.

EXTRACTION WITH WATER.

The determination of water-soluble constituents is made on the dried and weighed residues left in the crucibles after extraction with ether. The apparatus used consists of an ordinary heavy-walled side-neck suction flask provided with a rubber stopper, through which passes a carbon filter tube. The crucible is inserted in the top of the filter tube, a tight joint being obtained by means of a short length of thin-walled rubber tubing. As the analysis is made in duplicate, two suction flasks so arranged are connected to a Y tube, and both samples are extracted at once. A Bunsen valve or an empty bottle to serve as a trap should be inserted between the Y tube and the suction pump to guard against any tendency of the water to "suck back."

Cold water is used for the extraction because hot water would partly gelatinize any starch that might be present, and hot water would also remove more soluble organic material from the wood pulp. The water is passed through each sample in small quantities (about 20 c. c.) at a time. The sample is covered with water, allowed to stand a short time, and suction applied until all the water has passed into the flask. This process is repeated until at least 200 c. c. of water has been used. If each portion of water is allowed to stand on the sample for a short time and then thoroughly sucked out, this quantity of water is more than sufficient for complete extraction, but in case of doubt a few drops of the last portions of the filtrate is tested by evaporation on a glass plate.

If starch is present the filtration often proceeds very slowly because of the tendency of the starch to separate at the bottom of the crucible and form an almost impermeable layer on top of the asbestos mat. In such cases the use of stronger suction simply increases the density of this mass and retards rather than aids filtration. When any considerable quantity of starch has been detected in the qualitative examination, it is advisable to make use of porous alundum crucibles for the analysis, since these allow the filtrate to pass through the walls above the dense material at the bottom. With these crucibles it is necessary to use carbon tubes of such diameter that the crucible projects the greater part of its depth into the tube, being held by the rubber about one-fourth inch from its top. If this is not done there is a tendency for the filtrate to leak out of the crucible above the rubber.

These porous crucibles have been found decidedly convenient, especially in the case of materials that tend to clog the ordinary Gooch crucible.

The water extraction having been completed, the crucibles, with their contents, are again placed in the drying oven and dried for five hours at about 95 to 100° C. No additional loss results from longer drying at this temperature, and frequently to save time samples are dried overnight. After cooling in a desiccator the crucibles are weighed and the loss of weight noted. This loss of weight represents the total water-soluble material, and, in addition to the water-soluble salts detected in the qualitative examination, includes organic extract from the wood pulp, flour, or other absorbent. When cereal products are present the amount of organic material thus extracted may amount to 2 per cent or more, including sugars, etc., that form constituent parts of the grain. Frequently the antacid used, generally calcium carbonate or magnesium carbonate, is attacked by acid decomposition products from the nitroglycerin, a portion of the carbonate being thereby converted to nitrate or nitrite. In such cases some calcium or magnesium is found in the water extract.

Usually the only water-soluble constituent to be considered in an ordinary dynamite is an alkaline (sodium or potassium) nitrate. When an approximate analysis only is desired it is generally considered sufficient to regard the total loss of weight on extraction as nitrate, but, as shown above, this frequently gives erroneous results.

DETERMINATION OF ALKALINE NITRATES.

The method best suited for determination of nitrates is the following: An aliquot portion of the water extract is evaporated to dryness on a water bath and the residue gently ignited to burn off the organic matter. After cooling, the sides of the evaporating dish are washed down with a few cubic centimeters of water, about 1 c. c. of nitric acid is added, the evaporation repeated, and the residue heated gently over a burner until just fused, or the residue is dried in an oven at about 120° C. The treatment with nitric acid is necessary for the complete conversion to nitrate of any nitrite resulting from burning off the organic matter. The treatment should be repeated until the weight of the residue is constant.

The weighed residue is calculated as percentage of nitrate in the original explosive. Since this weight necessarily includes any non-volatile water-soluble impurities originally present in the nitrate, as iron, aluminum, chlorides, sulphates, etc., for an exact analysis it is necessary to ascertain the amount of such impurities by volumetric or gravimetric determinations on fresh aliquot portions of the water extract, or a direct determination of the true nitrate content may be made in the nitrometer as described on the following page.

DETERMINATION OF ALKALINE NITRATES BY MEANS OF THE NITROMETER.

An aliquot portion of the water extract estimated to contain the proper amount of nitrate for determination in the nitrometer (about 0.6 to 0.8 gram of $NaNO_3$ or 0.8 to 1.0 gram of KNO_3 for the type of nitrometer previously described, p. 35) is evaporated almost to dryness on a steam bath and transferred, by means of as little water as possible, to the cup of the nitrometer. The amount of water used should not exceed 20 c. c. This solution is drawn into the generator, and 30 to 40 c. c. of sulphuric acid (95 to 96 per cent) is added slowly, in small quantities at first to avoid generating sufficient heat to crack the glass. Because of the dilution of the acid the generation of the gas proceeds much more slowly than in the determination of nitroglycerin, and it is necessary to shake the generator for a total time of about 8 to 10 minutes in order to be certain that the reaction is complete. The volume of gas is measured and the percentage of nitrate is calculated in the same manner as in the case of nitroglycerin. This method is excellent for use as a check or for an exact determination of the actual nitrate content.

Tests made on a 1 per cent solution of pure potassium nitrate by both the gravimetric and volumetric methods described above gave results as follows:

Results of determinations of nitrates in a water solution by the gravimetric and by the volumetric method.

[Determinations by J. H. Hunter.]

Test No.	Volume of solution used.	Weight of KNO_3 used.	Gravimetric method. Weight of KNO_3 found.	Volumetric method. Reading of nitrometer (N).	Volumetric method. Weight of KNO_3 found.
	c. c.	*Grams.*	*Grams.*		*Grams.*
1	100	1.0000	0.9994	....	....
2	100	1.0000	....	13.92	1.0036
3	100	*a* 1.0000	.9997	....	....
4	100	*a* 1.0000	....	13.92	1.0036
5	100	*b* 1.0000	1.0004	....	....
6	100	*b* 1.0000	....	13.93	1.0043

a Two one-hundredths gram of sugar was added to the 100 c. c. of nitrate solution.
b Two one-hundredths gram of sodium chloride was added to the 100 c. c. of nitrate solution.

EXTRACTION WITH ACID.

As already pointed out, the materials most commonly used as antacids are the carbonates of calcium or magnesium or the oxide of zinc. Frequently ground dolomite is used, in which case both calcium and magnesium must be determined. The qualitative examination will have shown, however, what acid-soluble materials are present.

The procedure to be followed in making the acid extraction depends on whether or not starch is present in the explosive. In either case the dried and weighed residue insoluble in water is used for the treatment with acid.

In the absence of starch a simple extraction is made with cold dilute hydrochloric acid (1:10), 100 c. c. being drawn through the sample in the crucible in small successive portions, as described under "Extraction with water" (p. 43). Several portions of water are then drawn through to wash out the acid, and the crucible with the insoluble residue is dried as before for five hours at 95 to 100° C. It is sufficiently accurate to use this "loss-of-weight" figure as the amount of antacid, as the amount of organic material extracted from the wood pulp will be very small, but if greater accuracy is desired a quantitative determination of the dissolved base or bases may be made by the usual gravimetric methods

DETERMINATION OF CALCIUM.

Calcium is determined as follows: An excess of ammonium hydroxide is added and the solution boiled. Any precipitate of iron or aluminum hydroxides may be filtered off, ignited, and weighed, but the amount of such impurities is usually so small that it may be disregarded and calcium precipitated without previous filtration. Hot ammonium-oxalate solution is added in slight excess to the boiling solution and the boiling is continued for a short time. The precipitate is allowed to settle completely, and then is filtered, dried, and weighed as CaC_2O_4, or is ignited and weighed as CaO.

DETERMINATION OF MAGNESIUM.

Magnesium is determined in the filtrate from the calcium determination by concentrating to about 100 c. c., adding an excess of a solution of sodium hydrogen phosphate to the hot solution, then a large excess of ammonium hydroxide, and allowing the phosphate precipitate to separate completely by standing for several hours. The precipitate is filtered in a Gooch crucible, washed, ignited, and weighed as $Mg_2P_2O_7$.

DETERMINATION OF ZINC.

Zinc is precipitated with Na_2CO_3 solution as carbonate, ignited, and weighed as ZnO. If ammonium salts are present the zinc is precipitated with H_2S as ZnS, the ZnS filtered off, redissolved, and precipitated as carbonate. The determination of zinc is more fully considered in the discussion of ammonia dynamites on page 58.

DETERMINATION OF STARCH.

When starch is present both the starch and the antacid are removed in one operation by boiling with dilute acid, whereby the starch is rendered soluble by conversion to dextrin. In carrying out this process the material in the crucible is moistened with water and completely transferred with a spatula, or by washing with a stream of water from a wash bottle, into a beaker of about 500 c. c. capacity. If a Gooch crucible is used the asbestos is removed with the residue and the clean crucible dried and weighed. From the original weight of the crucible plus the asbestos the weight of the asbestos is ascertained and deducted from the final weight of dried residue remaining after hydrolysis. The volume of water in the beaker is made up to about 250 c. c., and about 3 c. c. hydrochloric acid (specific gravity 1.2) is added, and the mixture is stirred and brought to boiling over a burner. Boiling is continued until the starch is entirely hydrolized, a drop of the acid mixture being tested from time to time on a spot plate with a solution of iodine in potassium iodide until a blue coloration is no longer obtained. Longer boiling will only result in loss of extractive material from the wood pulp.

The boiled material is then at once filtered through a fresh crucible or through the original porous crucible, if such was used, washed several times with water, dried as before, and weighed. If the figures representing the weight include the weight of the asbestos mat from a Gooch crucible, the proper correction for the weight of the asbestos is made as noted above.

The amount of antacid contained in the acid filtrate is ascertained by a gravimetric determination as previously described.

It has already been noted (p. 44) that small amounts of soluble organic material (sugars, dextrin, etc.) from flour or other cereal products and extract from the wood pulp will be found in the water solution. In summarizing the results of analysis it is of course impossible to know what portion of such extracted organic material constituted part of the flour and what portion should properly be added to the wood pulp. Similarly, the organic material dissolved during the acid hydrolysis includes not only such portions of the grain as starch and gluten but soluble portions of the wood pulp.

It is customary in quoting the results of analysis to include all such soluble organic material from both water and acid extractions under the term "starch," and the insoluble residue is designated as "wood pulp and crude fiber." In other words, the weight of insoluble residue dried at 100° is called "wood pulp and crude fiber," whereas the sum of this constituent and of the ingredients determined in the ether, water, and acid solutions, deducted from the weight of original sample, is called "starch."

In the case of a dynamite which contains no cereal product, the sum of the determined ingredients deducted from the weight of original sample is taken as the amount of wood pulp present and includes the material extracted from the pulp by the water and acid treatment as well as the insoluble residue found by direct weight.

The actual amounts of wood pulp and of cereal products added in manufacture can not therefore be definitely determined, since portions of each will be found in the ether, the water, and the acid extractions as well as in the insoluble residue.

In order to determine to what extent the wood pulp in a dynamite is affected by the various extractions, etc., necessary in the course of analysis of the dynamite, dried samples of various grades of pulp were submitted to the treatment through which the insoluble wood pulp residue in a sample of dynamite had passed.

Samples of 2 to 3 grams of wood pulp were weighed in Gooch crucibles, dried to constant weight, and extracted successively with ether, water, and cold hydrochloric acid (1:10), and then boiled for 15 minutes with dilute hydrochloric acid (1:100). The latter treatment would be necessary if starch were present with the wood pulp in an explosive. After each operation the sample was dried five hours at 100°, and the loss of weight was determined. The results were as follows:

Results of analyses of wood pulp.

Sample No.	Loss of weight (per cent of dry sample.)				Percentage of insoluble residue.
	Extraction with ether.	Extraction with cold water.	Extraction with cold HCl(1:10).	Boiling with HCl (1:100)	
1.......	2.66	2.57	1.41	5.91	87.45
2.......	2.78	2.89	.42	1.75	92.16
3.......	2.09	2.23	.53	3.93	91.22
4.......	1.95	2.84	1.03	4.83	89.35

These experiments show that the final insoluble residue that is weighed as wood pulp may be only about 90 per cent of the amount of dry wood pulp actually present in the dynamite. A part of the loss is determined as rosin in the ether extract, and the portions extracted with water and hot acid are calculated as starch (if starch has been determined). When starch is not present the error in the determination of pulp is much less as the boiling-acid treatment is dispensed with. The analysis can then be made accurate by direct determination of the water-soluble nitrate and of the antacid as described above, and the amount of wood pulp can be found by subtracting the sum of the percentages of moisture, nitroglycerin, alkaline nitrate, and antacid from 100 per cent. This amount will be in excess of the percentage of insoluble residue found, according to the amounts of pulp extracted by the water and acid.

EXAMINATION OF INSOLUBLE RESIDUE.

DETERMINATION OF WOOD PULP, ETC.

The residue that remains after the ether extraction, the water extraction, and the extraction with dilute acid is usually a mixture of wood pulp, sawdust, or other form of cellulose or lignin. When corn meal, flour, middlings, bran, etc., are present in the dynamite, the residue will contain the nonstarchy portions of these materials, either free or mixed with wood pulp. In general, ordinary dynamite contains wood pulp alone as the absorbent, but low-freezing dynamites and "straight" dynamites containing less than 40 per cent of nitroglycerin often contain considerable quantities of corn meal, wheat middlings, or low-grade flour.

Infusorial earth was formerly much used as an absorbent for nitroglycerin, but in recent years it has seldom been so used in this country, having been almost entirely replaced by an active base or dope.

If the hydrochloric acid has not been thoroughly washed from the insoluble residue, the wood pulp will considerably darken in color during the drying process. From its physical structure, as observed without magnification, or with a small lens, much information may be gained in regard to the probable composition of the insoluble residue, but in all cases the examination is best made under the microscope, with a low-power objective, one of 32-mm. focus being suitable. One of the duplicate samples of insoluble residue is used for microscopic and chemical examination, and one for the determination of ash. A small amount of the sample which is to be used for microscopic and chemical examination is removed from the crucible, placed upon a microscope slide, and moistened with one or two drops of water. By means of a platinum needle the material is then carefully spread out, but no cover glass is used. Wood pulp, the most common constituent in the residue of ordinary dynamite, will be seen as separate fibers or bunches of fibers of very characteristic appearance.

In Plate IV, *A* and *B*, wood pulp of different grades is illustrated, and the characteristic appearance of sawdust or dust from certain types of woodworking machinery is shown in Plate IV, *C*. The bundles or clusters of fibers are characteristic of such materials. Typical samples of infusorial earth (kieselguhr) are shown in Plate IV, *D* and *E*. It should be noted that, as plainly shown in the figures, widely differing types of infusorial earth exist, forms of organisms appearing in one sample which will not be found in another. The general appearance of shell remains is a definite indication of the presence of diatomaceous or infusorial earth. The sample shown in Plate IV, *D*, was obtained, through the courtesy of Dr. G. P.

Merrill, from the National Museum, Washington, D. C. (Specimen No. 6555, from Cornwallis, Nova Scotia.)

The appearance of the "husks" or crude fiber from coarse wheat flour (middlings), after hydrolysis of the starch, is shown in Plate IV, *F;* the cellular structure of the irregularly shaped particles of fiber is readily seen.

A and *B*, Plate V, represent cellulose (cotton) and nitrocellulose, respectively. These two materials can not be distinguished from each other by microscopic examination in ordinary light, but in polarized light a decided difference is noted, the unnitrated fibers appearing in brilliant colors and the nitrated fibers dark.

In the case of the presence of cereal products it is often of value to make a microscopic examination of the residue insoluble in water before hydrolysis of the starch. In Plate V, *C, D,* and *E,* such materials are shown. *C* represents ordinary fine wheat flour; *D,* wheat flour mixed with wood pulp; *E,* coarse wheat flour or middlings; and *F,* corn meal. Characteristic differences in the appearance of the starch granules of wheat and corn are of aid in identifying these cereals, the wheat starch granules being in general well rounded or oval, whereas the cornstarch granules are almost always distinctly polygonal in shape.

DETERMINATION OF ASH.

The remaining sample of residue from acid extraction is used for the determination of ash, and may be either incinerated in the crucible that has been used for extraction, or the residue, together with the asbestos mat, may be removed to a platinum crucible and ignited; in this case there is subtracted from the ash the known weight of asbestos present in the mat. The ash of an ordinary dynamite, containing only wood pulp, sawdust, or corn meal as absorbents, will seldom amount to more than 0.20 per cent. When the amount of ash present is as much as 0.5 per cent the ash should in all cases be examined under the microscope to determine the possible presence of infusorial earth or other inorganic material (pulverized glass, sand, etc.), provided the presence of these materials has not already been detected by the microscopic examination. A high ash content may also indicate that either the water or acid extractions have not been complete. In this respect the ash determination may be regarded as a check on the analysis.

VARIATIONS DUE TO METHOD OF ANALYSIS.

In order to ascertain to what extent the results of analysis of a dynamite would be effected by variations in the method of analysis, uniform samples of a 45 per cent dynamite were submitted to the laboratories of 11 explosives works, with the request that analyses be

A. WOOD PULP NO. 1 (× 50).

B. WOOD PULP NO. 2 (× 50).

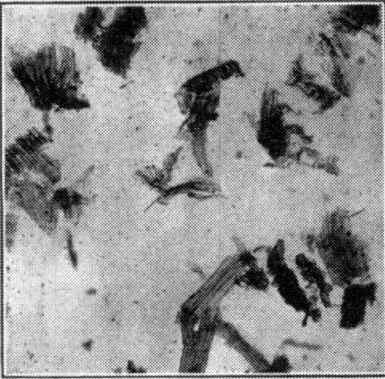

C. SAWDUST (× 25).

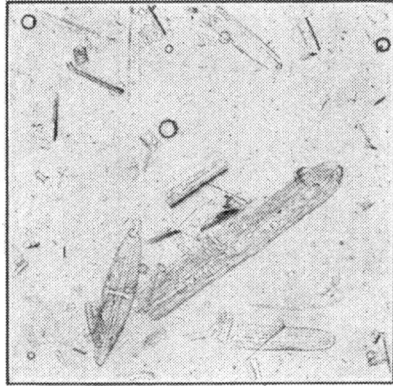

D. INFUSORIAL EARTH NO. 1 (× 150).

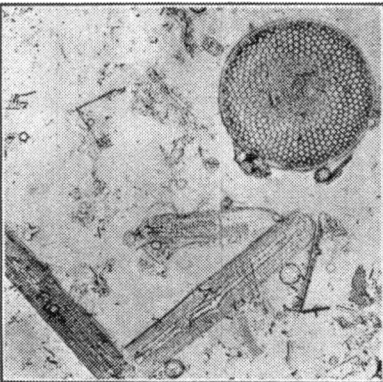

E. INFUSORIAL EARTH NO. 2 (× 150).

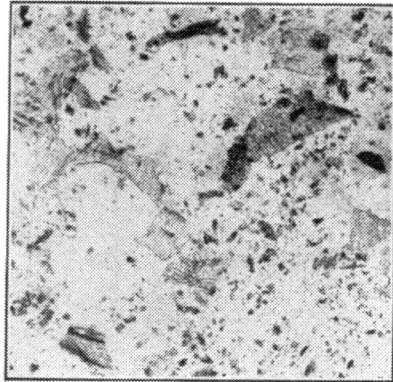

F. CRUDE FIBER FROM WHEAT MID-
DLINGS (× 25)

made by the methods regularly in use in each laboratory and results reported, together with brief notes as to the method used.

The samples were prepared as follows: Fifty cartridges of a lot of 45 per cent dynamite were opened, about two-thirds of each cartridge removed (the ends being rejected), and broken up finely in a porcelain dish by means of a horn spoon. All of these portions were then mixed together very thoroughly in a large porcelain dish. From this uniform mixture about 20 sample bottles (150 c. c. capacity) were filled, the contents of each bottle then being emptied out separately into a dish, carefully mixed again, and replaced in the bottle. Each sample represented about 150 grams.

Eleven of these samples were sent to different laboratories, as noted above, and seven of them analyzed in the bureau's explosives chemical laboratory by different analysts. Careful instructions were sent with each sample, that, in order to compensate for any segregation occurring in shipment, the entire sample should be thoroughly mixed before analysis.

The results of the analyses of these samples are shown in the following table:

Analyses of uniform samples of dynamite.

1	2	3	4	5	6	7	
						Nitroglycerin.	
Sample No.	Moisture.	Nitro-glycerin.	Potas-sium ni-trate.	Calcium carbon-ate.	Wood pulp.	By direct weighing.	By ni-trometer.
A	1.21	43.90	39.14	1.02	14.73	....	....
B	.89	44.99	38.40	1.15	14.57	....	....
C	.99	44.77	38.75	1.06	14.59	....	....
D	.91	45.25	38.62	.98	14.24	....	....
E	1.05	44.52	38.05	1.28	15.10	....	....
F a	1.02	42.29	41.02	1.15	14.52	....	....
G	.41	44.91	39.04	1.12	14.52	....	....
H	1.35	44.86	38.09	1.28	14.42	....	....
J	1.22	44.98	38.48	.99	14.37	....	....
K b	....						
L	1.60	44.50	37.40	1.00	15.50	....	....
M	1.00	45.70	37.57	1.17	14.56	....	....
N	1.19	45.50	37.54	1.15	14.62	....	....
O	1.15	45.48	37.69	1.16	14.52	45.87	45.22
P1	1.04	45.35	37.66	1.10	14.85	45.31	45.17
P2	1.26	45.22	37.33	1.20	14.88	....	....
P3	.98	45.24	37.87	1.21	14.70	....	....
Q	1.18	45.42	37.60	1.19	14.61	45.41	45.14

a The remarkable difference noted in the analysis of this sample as compared with all other samples, is due to the fact that, through misunderstanding, this sample was not mixed on being received. After several analyses had been made with widely varying results, the remainder of the sample was mixed by rubbing through sieves. This treatment probably resulted in considerable loss of nitroglycerin. Sample F is therefore not considered in the discussion of results.

b No report received.

Samples A to L were analyzed in the laboratories of various powder companies and samples M to Q in the bureau's laboratory; P2 and P3 were analyzed by men under instruction, not members of the laboratory force of the bureau.

DISCUSSION OF ANALYSES.

MOISTURE.

The methods employed were as follows:

Samples E, J, and M to Q, inclusive—3 grams desiccated on watch glass over H_2SO_4 for 3 days (0.98–1.26 per cent).

Sample A—6 to 10 grams on watch glass, 3 days over H_2SO_4 (1.21 per cent).

Sample B—3 grams on watch glass, 2 days over $CaCl_2$ (0.89 per cent).

Sample D—6 grams on watch glass, 24 hours over H_2SO_4 in vacuum (0.91 per cent).

Sample C—2 grams on watch glass, 24 hours over H_2SO_4, 15 inches vacuum (0.99 per cent).

Sample H—5 grams in Gooch crucible, 48 hours over H_2SO_4 at 40° C. (1.35 per cent).

Sample G—10 grams in 4-ounce bottle, dry-air current passing *over surface* of sample for 24 hours (0.41 per cent).

Sample L—10 grams in drying tube, dry-air current passing *through* sample for 24 hours (1.60 per cent).

It is noted that desiccation on a watch glass over H_2SO_4 for three days gave only 0.3 per cent maximum variation, results obtained with the aid of vacuum for 24 hours being a little low. Two days over $CaCl_2$ gave low results. High results were obtained by desiccating at 40° C. and by passing dry air through the sample, from loss of nitroglycerin under these conditions. Passing dry air over the surface of the sample gave low results, as might be expected.

NITROGLYCERIN.

The results obtained in the bureau's laboratory with samples M to Q varied from 45.22 to 45.70 per cent. Samples of 5 to 10 grams in Gooch crucibles were extracted for one hour with U. S. P. ether (96 per cent) in the Wiley apparatus and the residues dried five hours at 100° C. The loss in weight minus the moisture previously determined was taken as the nitroglycerin content. Samples B, E, and J were analyzed in the same manner, giving results from 44.52 to 44.99 per cent. Samples A, C, D, G, and L were extracted with ether by means of suction, 7 to 10 grams of sample being treated with five or six successive portions of ether (amounts varying from 50 to 120 c. c.). Samples A, C, and D were dried to constant weight in steam ovens after extracting (the time of drying not noted); G was dried at 100° to 105°, and L for two hours at 95°. The results varied from 43.90 to 45.25 per cent. The results obtained in the bureau's laboratory (45.22 and 45.70 per cent) are uniformly higher than those obtained in other laboratories (43.90 to 45.25 per cent).

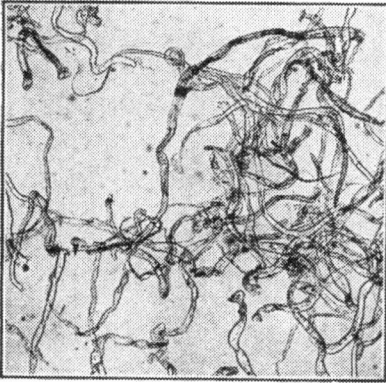

A. CELLULOSE (COTTON) (× 50).

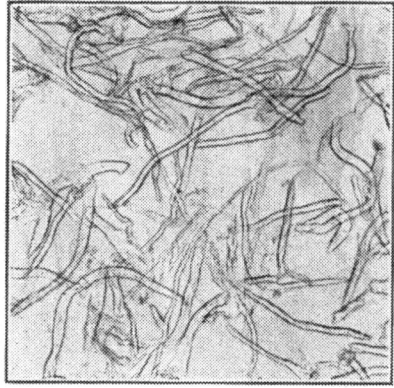

B. NITROCELLULOSE (× 50).

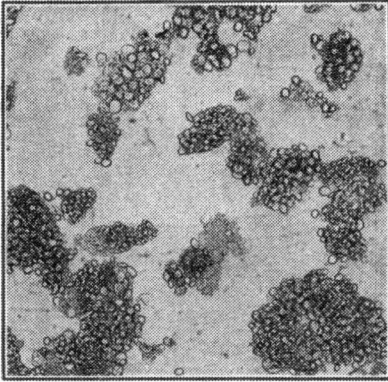

C. WHEAT FLOUR (FINE) (× 50)

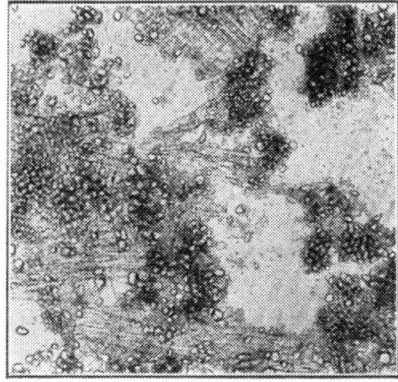

D. WHEAT FLOUR (FINE) AND WOOD PULP (× 50).

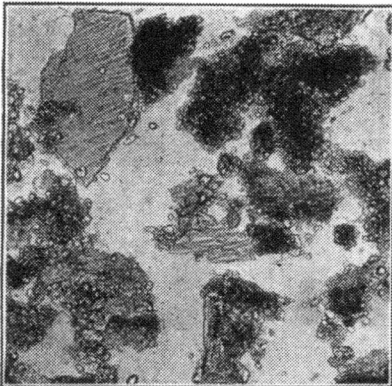

E. WHEAT FLOUR (MIDDLINGS) (× 50).

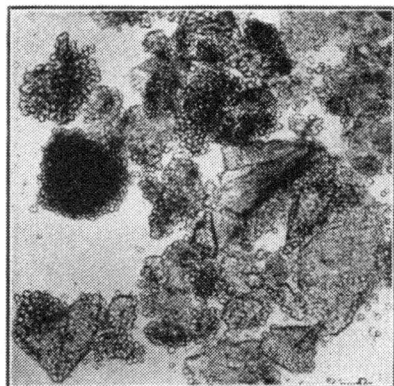

F. CORN MEAL (× 50).

Nitrometer determinations on three samples (column 7) showed the true nitroglycerin content of the evaporated ether extracts to be approximately 45.2 per cent.

No explanation can be given for the large number of results in which the amounts of nitroglycerin found are uniformly low, except failure to observe one or more of the following precautions: (1) The extraction must be complete; (2) the residue must be dried to constant weight at a temperature of about 100° C.; (3) the dried residue must be cooled in an efficient desiccator and weighed as soon as cooled.

POTASSIUM NITRATE.

The results obtained in the bureau's laboratory varied from 37.33 to 37.87 per cent. The determination was made by extracting the residue insoluble in ether with about 200 c. c. of water, and determining the nitrate in an aliquot part of the water solution by evaporation, as described on page 44. No correction was made for traces of chlorides, etc., in the solution. The determination on sample H was made in the same manner, giving 38.09 per cent, while on sample L an aliquot portion of the water extract was analyzed in the nitrometer with a result of 37.40 per cent. All the remaining samples were extracted with water, the insoluble residues dried at about 100 C°. and weighed, the loss of weight being regarded as potassium nitrate. This method gave high results (38.05 to 39.14 per cent) owing to the water-soluble organic matter extracted from the wood pulp.

CALCIUM CARBONATE.

In samples L to Q the calcium carbonate was determined gravimetrically in the dilute-acid extract; in samples C and D by direct titration of the residue insoluble in water, or of the ash left after burning off the pulp; in A, B, E, and J the loss of weight on extraction with dilute acid was considered as calcium carbonate, while in G and H the ash was assumed to be entirely calcium carbonate. The variations are of minor importance and no conclusions can be drawn from the results.

WOOD PULP.

Variations in the method of determining the potassium nitrate influence the proportion of wood pulp reported. When the loss of weight on extracting with water is assumed to be entirely potassium nitrate, in spite of the fact that it contains considerable organic matter extracted with the wood pulp, the proportion of wood pulp found will be lower than if the nitrate is determined by a direct method and the wood pulp by difference. The percentage of wood pulp being found by subtracting the sum of all other constituents from 100 per cent, no comparison of results is possible.

GELATIN DYNAMITE.

Unlike ordinary dynamite, which contains nitroglycerin absorbed in a porous material, gelatin dynamite contains nitroglycerin combined with nitrocellulose to form a plastic solid. When nitroglycerin is warmed with nitrocellulose containing about 12 per cent of nitrogen, the nitroglycerin dissolves the nitrocellulose, and a thick, viscous mass is produced which resumes a jelly-like consistency as soon as it has cooled. When as little as $3\frac{1}{2}$ per cent of nitrocotton is dissolved in nitroglycerin at 60°, the material should form a jelly-like nonflowing mass when cooled to ordinary temperatures, but when smaller amounts of nitrocellulose are used the viscosity of the solution becomes less. The explosive known as "blasting gelatin" consists of about 93 to 90 per cent of nitroglycerin and 7 to 10 per cent of nitrocellulose, and is a translucent, jelly-like mass, containing the highest percentage of nitroglycerin used in any solid explosive.

Any explosive containing nitroglycerin combined with nitrocellulose in connection with an active base consisting of a nitrate and combustible material is termed a "gelatin dynamite." The gelatin dynamites have many properties that make them desirable for mining work, their greatest advantage being that they are almost unaffected by water.

SAMPLING.

The coherent, pasty, doughlike consistency of gelatin dynamite renders the preparation of a uniform sample much more difficult than is the case with ordinary dynamite.

A sample is prepared from one or more cartridges by cutting off portions from different parts of each stick; these portions are then cut into thin pieces and broken up as finely as possible by means of an aluminum or platinum spatula. The use of a steel spatula or knife is not to be recommended. The sample so prepared is well mixed and bottled, and because of its tendency to form a solid mass again on standing, it should be analyzed as soon as possible after being prepared.

The ingredients that may be found in the various types of gelatin dynamite are nitroglycerin; nitrocellulose; sulphur; rosin; sodium, potassium, or ammonium nitrate; calcium or magnesium carbonate; wood pulp; and cereal products.

Moisture is determined in the manner previously described.

54

The extraction with ether is made as for dynamite except that ether distilled over sodium (that is, ether free from alcohol) is used in order to avoid the partial solution of the nitrocellulose. Nitrocellulose is insoluble in pure ether, but a small percentage of alcohol present as an impurity may cause the solution of a considerable proportion of this constituent, and as the amount of nitrocellulose is usually only 0.5 to 2 per cent its determination should be accurate.

The ether solution containing the nitroglycerin, sulphur, and rosin is treated in the manner already described, and the water extraction of the dried and weighed insoluble residue is made in the usual way.

SULPHUR.

If the percentage of sulphur is unusually high, or if the extraction with ether has not been continued long enough, some sulphur may remain in the dried residue left after water extraction, in which case an additional extraction is made with carbon disulphide in the Wiley apparatus, the same method as described for the ether extraction being used. After the extraction with carbon disulphide has been made, the crucibles should be sucked dry and the carbon disulphide allowed to evaporate in a warm place before the crucibles are placed in the oven, as the vapors of carbon disulphide are very inflammable and may ignite in the oven. The crucibles with their dried residue are weighed and the loss of weight considered as sulphur.

The extraction with water is next made as before described.

NITROCELLULOSE.

Nitrocellulose is now removed by means of a suitable solvent. The pyroxylin cotton usually employed as a gelatinizing agent is soluble in a mixture of two parts ether and one part alcohol, but as all grades of nitrocellulose are more readily soluble in acetone than in ether alcohol it is customary to use acetone as the solvent.

The extraction with acetone is made by separating the dry residue from the crucible, leaving the mat intact if possible, placing the residue in a small beaker and covering it with acetone. The mixture is allowed to stand for three to four hours, with frequent stirring to dissolve completely the nitrocellulose, and is then filtered through the original crucible, washed with acetone, dried in the usual manner, and weighed. The loss of weight represents nitrocellulose plus a small amount of extract from the wood pulp. The wood pulp extract is usually so small that it may be disregarded, but a check on the nitrocellulose determination may be made by evaporating the acetone solution to a small volume (about 20 c. c.), and diluting gradually with a large volume of hot water (about 100 c. c.), which drives off the volatile solvent, precipitating the nitrocellulose as a white flocculent mass. The precipitate is then filtered off, dried, and weighed as nitrocellulose.

The remainder of the analysis is conducted as is that described for straight dynamite.

Some of the older types of gelatin dynamites contained small amounts (1 to 2 per cent) of paraffin, but this is an unusual ingredient and is more frequently found in ammonia dynamite. (See p. 57.)

Ammonia gelatin dynamite is a type that has of recent years assumed commercial importance. It differs from ordinary gelatin dynamite in the fact that it contains ammonium nitrate in addition to the usual constituents of the former. As in the ammonia dynamites, discussed later, so here the ammonium nitrate is usually previously coated with vaseline, paraffin, or other waterproofing material, and is neutralized with zinc oxide.

In the analysis of gelatin dynamite it should be remembered that trade custom has led to an erroneous system of designating the strength of explosives of this class. Thus a gelatin dynamite containing about 30 to 33 per cent of nitroglycerin and about 1 per cent of nitrocotton is called a "40 per cent" strength gelatin dynamite. This unfortunate practice undoubtedly had its origin in the fact that, as gelatin dynamite is much denser than ordinary dynamite, and a greater quantity can therefore be placed in a hole, it was assumed to be stronger. Comparative tests indicate that, weight for weight, a so-called "40 per cent" strength gelatin dynamite containing 33 per cent nitroglycerin is much weaker than is an ordinary "straight" dynamite containing 40 per cent nitroglycerin.

AMMONIA DYNAMITE.

The usual type of ammonia dynamite is practically a "straight" dynamite in which a large part of the nitroglycerin is replaced by ammonium nitrate. Sulphur is sometimes a constituent of this type of explosive, and frequently the wood pulp is wholly or largely replaced by coarse flour or middlings. The ammonium nitrate is generally protected from hygroscopic influence by a coating of vaseline or paraffin and is neutralized with zinc oxide. These ingredients are added to the ammonium nitrate in the course of its manufacture while the crystals of the ammonium nitrate are still hot.

In the analysis of such explosives the determination of moisture and extractions with ether, water, and acid are carried out as previously described. An additional extraction with carbon disulphide is usually necessary in order to remove all of the sulphur; this is done after the water-soluble salts have been extracted.

One portion of the ether extract is used for the determination of nitroglycerin in the nitrometer and the other for the determination of the sulphur and vaseline or paraffin.

The method for the analysis of the ether extract as described on pages 41 and 42 is the scheme of separation followed when both sulphur and vaseline, or paraffin, are present with the nitroglycerin, although several other methods are applicable and give reliable results.

The nitroglycerin may be destroyed by means of caustic alkali, which also dissolves any resin present; the solution is decanted from the residue of paraffin, or vaseline, and sulphur; the resin is precipitated with hydrochloric acid, filtered, dried, and weighed. The residue of sulphur and paraffin or vaseline is treated with hot ammonium sulphide which dissolves the sulphur; this solution is cooled, decanted, and the vaseline or paraffin adhering to the beaker is washed, dried, and weighed. The weight of sulphur is found by difference.[a]

The water extract contains both ammonium and sodium nitrates, together with water-soluble organic material from the flour or other absorbent. If zinc oxide has been used as the antacid, all of this component will generally be found in the water extract, since the small amounts used are readily soluble in ammonium nitrate solutions.

[a] Gody, L., Traité théorique et pratique des matières explosives, 1907, pp. 388-389.

An aliquot part of the water extract is evaporated in a platinum dish on a steam bath, the ammonium nitrate volatilized, and the organic matter burned off by careful heating over a burner. A little nitric acid is added to oxidize to nitrate any nitrite resulting from reduction, as described on page 44. In heating this residue care must be taken to avoid decomposition of the zinc nitrate, or else the heating should be strong enough to convert it entirely to zinc oxide. Either of the following methods may be used:

(1) The residue, after evaporation of the nitric acid on the steam bath, is dried at about 110° to 120°, and weighed as $NaNO_3$ and $Zn(NO_3)_2$. This residue is then dissolved in water and the zinc precipitated with sodium carbonate, filtered, ignited, and weighed as ZnO. The weight of $NaNO_3$ and $Zn(NO_3)_2$ minus (2.33 times the weight of ZnO) equals $NaNO_3$.

(2) The residue obtained as above is heated gently over a burner until the evolution of oxides of nitrogen from the decomposition of the zinc nitrate has ceased, care being taken that the temperature is not high enough to cause a loss of sodium nitrate. The residue is now weighed as $NaNO_3$ plus ZnO, then treated with water to dissolve the $NaNO_3$; the ZnO is filtered on a Gooch crucible, ignited, and weighed as ZnO, the $NaNO_3$ being found by difference from the combined weight. The filtrate containing the sodium nitrate should be tested with ammonium sulphide to assure that the zinc has been entirely converted to insoluble zinc oxide.

Zinc may be determined in a separate portion of the water extract by adding ammonia, precipitating with hydrogen sulphide, and filtering off the precipitated zinc sulphide. The precipitate is washed and, without drying, dissolved in a small amount of nitric acid, and evaporated to dryness. Any free sulphur is thus oxidized. The residue is treated with a little sulphuric acid and again evaporated to dryness over a burner at a dull-red heat until the free acid has been volatilized. Little, if any, decomposition of the zinc sulphate results from this heating. The treatment with sulphuric acid and heating should be repeated until a constant weight of $ZnSO_4$ is obtained. This method is convenient and has been found to check well with the methods described above.

Ammonium nitrate is determined directly with a separate portion of the water extract by the usual method of distilling from a solution made strongly alkaline with KOH, collecting the distillate in a known volume of standard H_2SO_4, and titrating the excess of acid with standard alkali, cochineal being used as an indicator.

In regard to the determination of ammonium nitrate, the possible effects of several influencing factors have been investigated. Stillman and Austin state that ammonium nitrate is slightly soluble in ether,[a]

a Kast, H.. Anleitung zur chemischen und physikalischen Untersuchungen der Spreng- und Zündstoffe, 1909, p. 980.

and several other authors have noted that correction should be made for this solubility.

Ten-gram samples of pure ammonium nitrate were desiccated to constant weight over sulphuric acid and were extracted with U. S. P. (96 per cent) ether for one hour in the Wiley apparatus. On drying to constant weight in vacuum desiccators a maximum loss of only 0.10 per cent was noted. Additional drying of the same samples at 70° for 18 hours caused a further loss of 0.07 to 0.15 per cent. On continuing the drying at 90° to 100° further losses calculated as percentages of the original sample were as follows:

Loss in weight of ammonium nitrate dried at 90° to 100°.

Time of drying.	Total loss, per cent.	
	Sample 1.	Sample 2.
Hours.		
2	0.03	0.04
5	.03	.05
24	.07	.11
120	.33	.39
240	.62	.67

The above-mentioned experiments show that the loss of pure ammonium nitrate by the ether extraction and by subsequent drying for several hours at 70° or 100° is very slight—not over 0.25 per cent.

That the loss on drying depends on the surface exposed was shown by drying, at 90° to 100°, 5-gram samples spread on 3-inch watch glasses. The losses were as follows:

Loss in weight of ammonium nitrate dried on 3-inch watch glasses at 90° to 100° C.

Time of drying.	Loss of weight, per cent.	
	Sample 1.	Sample 2.
Hours.		
5	0.34	0.33
24	.88	.98
48	2.03	1.57
120	6.11	4.64
168	7.09	5.84

In this case the much greater surface of samples exposed caused a much more rapid volatilization.

In the analysis of an explosive containing zinc oxide, it must be remembered that in addition to mere volatilization there is possible still further loss from the decomposing action of the ZnO on the ammonium nitrate. This is shown by the following experiment: Two 10-gram samples of ammonium nitrate were ground in a mortar, 0.5 gram zinc oxide being well mixed with one sample. The two samples were then placed in crucibles and heated at 100°. The losses in weight are given in the table following:

Loss in weight of ammonium nitrate with and without zinc oxide on heating at 100°.

Time.	Loss in weight, per cent.	
	Ammonium nitrate.	Ammonium nitrate plus 5 per cent of zinc oxide.
Hours.		
2	0.02	0.61
18	.11	1.07
48	.16	1.13

The effect of too high temperature of drying is shown more clearly in the following experiments with an ammonium nitrate explosive of the "permissible" class, containing nitroglycerin. The explosive contained about 75 per cent of NH_4NO_3 and 1.4 per cent of ZnO. Six 10-gram samples were extracted with ether in the usual manner to remove the nitroglycerin. Two of the extracted samples were dried over night at 100° C., two over night at 70° C., and two for 24 hours in vacuum desiccators to constant weight. The samples were then extracted with water, dried five hours at 100° C., and the loss in weight noted. Ammonium nitrate was determined in the water solutions by the distillation method previously described. The results are tabulated below:

Effect of different methods of drying ammonium nitrate explosives after ether extraction.

[Determinations by J. H. Hunter.]

Method of drying.	Loss of weight, per cent.						NH_4NO_3 in water solution, per cent.		
	Ether extraction.			Water extraction.					
	High.	Low.	Mean.	High.	Low.	Mean.	High.	Low.	Mean.
In a 100° oven..............	13.48	13.10	13.29	77.58	77.23	77.40	....	71.76	71.76
In a 70° oven...............	11.72	11.64	11.68	79.29	79.09	79.19	75.01	74.78	74.90
In a vacuum desiccator......	11.68	11.57	11.63	79.22	79.11	79.17	75.31	75.20	75.25

These results show clearly that drying at 100° after extraction with ether causes a considerable loss of ammonium nitrate, most of the loss being due to decomposition of the nitrate by the zinc oxide.

The sum of the amounts of sodium nitrate, ammonium nitrate, and zinc oxide found in the water solution will be less than the total water extract, since the latter, as before noted, includes water-soluble organic material from the wood pulp, flour, or other carbonaceous absorbents. This organic material is added to the amount of starch, etc., removed by hydrolysis and reported as "starch," as has been noted on a previous page.

LOW-FREEZING DYNAMITE.

Nitroglycerin congeals or freezes at a temperature of about 8° C. (46° F.), and as nitroglycerin in the frozen condition is much less sensitive to the action of a detonator than when in the liquid condition, explosives containing frozen nitroglycerin usually fail to explode when attempts are made to use them.

As the outdoor temperature for several months in the year is on the average below the temperature at which nitroglycerin explosives freeze, many efforts have been made to prepare low-freezing explosives, which, without being thawed, may be used at temperatures lower than that at which ordinary nitroglycerin explosives are frozen. Some types of nonfreezing explosives contain ammonium nitrate mixed with nitrostarch, nitrotoluene, or similar material, but those explosives to which the term "low-freezing" is usually applied contain nitroglycerin mixed with the liquid nitrotoluenes, with crystalline dinitrotoluene or trinitrotoluene, with the nitrochlorhydrins or other materials, it having been shown, for example, that 20 per cent of dinitrochlorhydrin in nitroglycerin reduces the temperature at which the mixture will freeze to about —12° C.,[a] and by the use of mixtures of nitroglycerin and dinitrotoluene explosives are prepared which may be used at temperatures considerably below 0° C.

Many low-freezing explosives are marked with the designation "L. F." upon their wrappers; but even in the absence of this notice of the nature of the explosive a low-freezing dynamite may frequently be distinguished by the odor of nitrotoluene, or by the color test for nitrosubstitution compounds made on the ether extract as described under "Qualitative examination," on pages 17 and 18.

DETERMINATION OF NITROSUBSTITUTION COMPOUNDS.

With low-freezing dynamites the methods of sampling and of determination of moisture are carried out in the manner already outlined. The extraction with ether is also made in the usual manner, and one of the samples of ether extract is used, as outlined on a preceding page, for the determination of resins, sulphur, or other ingredients. The other sample of ether extract is used for the determination of the nitroglycerin and nitrosubstitution products.

---

[a] Roewer, F. A., Proc. 6th Int. Cong. App. Chem., vol. 1, 1906, p. 541.

So far as is known by the writers, no satisfactory method has yet been found for the direct determination of nitrosubstitution compounds in the presence of nitroglycerin. The analysis of such mixtures is generally made by determining the nitroglycerin by means of the nitrometer (p. 35), and finding the amount of nitrosubstitution compound by difference. The nitrosubstitution compounds are not decomposed in the nitrometer as are the nitric esters.

It is interesting to note, however, that if mononitrotoluene is present the amount of nitroglycerin found by means of the nitrometer does not represent the true amount of this ingredient in the mixture. It has been found [a] that a portion of the nitric acid liberated from the nitroglycerin by the action of the sulphuric acid used in the determination is taken up by the mononitrotoluene present, the latter becoming quantitatively nitrated to dinitrotoluene. Thus, if the amount of nitroglycerin in the ether extract is not in excess of the theoretical amount required to yield sufficient nitrogen to convert all the mononitrotoluene to dinitrotoluene, no evolution of nitric oxide will result. In other words, the error in the determination of nitroglycerin is equal to about 0.5530 gram of nitroglycerin for each gram of mononitrotoluene present. Pure crystalline dinitrotoluene or trinitrotoluene was found to have no effect on the determination, and the same was found to be true of the so-called "liquid trinitrotoluene." "Liquid dinitrotoluene," however, causes an error amounting to 0.0628 gram of nitroglycerin for each gram of the nitrosubstitution product. This error was shown to be probably due to the presence of mononitrotoluene in the liquid dinitrotoluene.

These results show that the nitrometer method for determination of nitroglycerin is not reliable if mononitrotoluene is present. The latter is, however, seldom used in low-freezing dynamites, and in the case of the more commonly used liquid dinitrotoluene, the resulting error is not large. For example, in an explosive containing 25 per cent of nitroglycerin and 10 per cent of liquid dinitrotoluene, the error in the determination of nitroglycerin would amount to 0.628 per cent, or the amount of nitroglycerin found would be 24.37 per cent instead of 25 per cent.

Further information may be gained by determining the total nitrogen of both the nitroglycerin and the nitrosubstitution compound by a modification of the Kjeldahl method and deducting the nitrogen of the nitroglycerin, the difference being the nitrogen of the nitrosubstitution compound. From this nitrogen the amount of the latter can be calculated. This method is of value only if the

a Storm, C. G., The effect of nitrotoluenes on the determination of nitroglycerin by means of the nitrometer, Proc. 8th Int. Cong. App. Chem., vol. 4, 1912, p. 117.

nitrosubstitution compound has been identified by the preliminary examination.

Another method for the determination of the total nitrogen of such mixtures, as well as the ester nitrogen, is that of Berl and Jurrissen,[a] in which a so-called "decomposition flask" is used. By means of sulphuric acid and a small amount of mercury in a vacuum the decomposition of nitroglycerin or other nitric ester is effected and the resulting volume of nitrogen oxide (NO) measured. A second sample of the mixture is treated in the same manner after a preliminary oxidation with chromic acid and sulphuric acid, whereby the total nitrogen is converted to nitrogen oxide (NO). The difference between the two volumes of nitrogen oxide (NO) equals that resulting from the nitrosubstitution compound.

The remainder of the analysis—the extractions with water, acid, etc.—is conducted in the manner previously described and should present no further difficulties.

---

[a] Berl, E., and Jurrissen, A. W., Über gasvolumetrische Analyse mit dem "Zersetzungskolben" und die Stickstoffbestimmung in rauchschwachen Pulvern. Zeitschr. angew. Chem., vol. 23, 1910, p. 241.

GRANULATED NITROGLYCERIN POWDER.

When ordinary dynamite is used in blasting any material of a consistency resembling that of earth or clay, little of the energy of the explosive does useful work, because the soft and compressible nature of the material blasted allows the greater part of the energy of the explosive to be used in compacting the material and in producing a cavity. Consequently, gunpowder has been largely used in all places where earth, clay, or other soft material was to be dislodged. In the blasting of the banks of railroad cuts there are often places where a soft, but somewhat consolidated material, intermediate between earth and hard rock, has to be blasted. Such a material might be a soft shale, for example, or a friable and easily crumbled sandstone, and for dislodging it neither ordinary blasting powder nor ordinary dynamite is particularly suitable. In the year 1876 E. Judson patented an explosive consisting of a low-grade gunpowder made by heating and mixing together coal, sulphur, sodium nitrate, etc., granulating the mixture, and then coating this non-absorbent granulated dope with a small amount of nitroglycerin. The proportion of nitroglycerin used—often as little as 5 per cent— was such that had it been absorbed by the grains of explosive it would not have been capable of detonation, but by remaining wholly or largely upon the surface of the grains the use of a detonator brought about its explosion and the simultaneous ignition of the gunpowder base which it covered. Such an explosive, as a result of the detonation of the nitroglycerin, produces an initial blow sufficient to crack and fissure the partly consolidated material in which it is placed. The action of the gunpowder mixture that forms the larger part of the explosive so heaves and moves the broken-up mass as to make easy its removal with steam shovels.

Granulated nitroglycerin powders, or "free-running" explosives, have been very much used in the excavation of earth and are commonly known as Judson powder (after the inventor), bank powder, or railroad powder. In the analysis of low-grade granular powder, moisture is determined by the standard method, and the usual method of extracting with ether is followed. In the ether extract are usually found large proportions of sulphur, rosin, etc., besides the nitroglycerin. The proportion of sulphur commonly used in

64

low-grade granular powder is so considerable that usually it is not all removed by extraction with ether.

The determination of nitrates in the residue after ether extraction is made in the manner already outlined for ordinary dynamite; an additional extraction with carbon disulphide is necessary to remove the sulphur not extracted by means of ether.

As antacids are seldom added to low-grade granular explosives, the extraction with dilute acid may generally be omitted; but if the qualitative examination has indicated the presence of an antacid, its determination is made as already described.

The residue remaining after the extractions with ether, water, and carbon disulphide is usually bituminous coal, although charcoal or other carbonaceous material may be found. An examination under the microscope or with a hand magnifier will usually show with sufficient certainty the nature of the insoluble residue, and the presence of bituminous coal may generally be confirmed by a volatile-matter determination made by heating the solid residue in a small crucible over a Bunsen flame.

The separation and determination of the ingredients of the ether and water extracts is carried out by the methods previously described.

67709°—Bull. 51—13——5

BLACK POWDER.

The general term "black powder" is applied to several explosives of nearly similar composition, including chiefly black blasting powder, black gunpowder, and black fuse powder. As their chemical examination involves identical problems, they are here treated as one general class.

Black blasting powder usually consists of a mixture of sodium nitrate, sulphur, and charcoal, whereas black gunpowder is generally a mixture of potassium nitrate, sulphur, and charcoal. The real difference between "gunpowder" and black blasting powder is one of use, since some blasting powder containing potassium nitrate as the oxidizing material has been made, and, similarly, there is record of gunpowder having been made from sodium nitrate. Although black blasting powder is usually made in larger grains than gunpowder, yet for certain purposes, particularly for large cannon, grains of gunpowder have been made even larger in size than the customary kinds of blasting powder, and again finely granulated blasting powder has been made for use where a quick-acting explosive, yet one not so rapid as dynamite, was desired.

The black-powder composition used in the ordinary miner's safety fuse and known as "fuse powder" is much similar to the ordinary grade of gunpowder, but is of very fine granulation (usually 40 to 100 mesh), and contains potassium nitrate as its oxidizing agent.

PHYSICAL EXAMINATION.

GRANULATION OR AVERAGE SIZE OF GRAINS.

The determination of the granulation of black powder is made by a series of standard sieves, but this examination is not usually required in connection with chemical analysis, and can be made only where a large sample of the powder is available. The standard sizes of grains of black blasting powder, together with a statement of the size of screen through which the material will pass, is given in the following tabulation:[a]

---

[a] Munroe, C. E., and Hall, Clarence, A primer on explosives for coal miners. Bureau of Mines, Bull. 17, 1911, p. 17.

Relation between sizes of black blasting powder and separating sieve.

Size of grains.	Diameter of round holes in screens through which grains pass.	Diameter of round holes in screens on which grains collect.
CCC	$\frac{40}{64}$ inch	$\frac{32}{64}$ inch
CC	$\frac{36}{64}$ inch	$\frac{24}{64}$ inch
C	$\frac{27}{64}$ inch	$\frac{18}{64}$ inch
F	$\frac{20}{64}$ inch	$\frac{12}{64}$ inch
FF	$\frac{14}{64}$ inch	$\frac{7}{64}$ inch
FFF	$\frac{9}{64}$ inch	$\frac{3}{64}$ inch
FFFF	$\frac{5}{64}$ inch	(*a*) $\frac{2}{64}$ inch

a Or 28-mesh bolting cloth.

GRAVIMETRIC DENSITY.

By "gravimetric density" is meant the "apparent specific gravity" of the explosive, or the ratio that the weight of the powder contained in a given volume bears to the weight of water that would exactly fill the same volume. Gravimetric density or apparent specific gravity is therefore not only a factor of the true density of the powder, but is also influenced by the size and the shape of the grains, since obviously the space occupied by voids, or the space between grains, must vary with the shape and size of the grains.

The standard determination of gravimetric density is made by pouring the powder into a vessel, usually in the shape of the frustrum of a cone, striking off with a straightedge all over that required to fill the measure, and then weighing the powder held by the receiver. Plate II, *B*, shows a commercial type of gravimetric balance for this purpose. With this balance direct readings of the gravimetric density of a powder are possible without calculation. When so few determinations must be made as to make the use of a separate instrument seem unnecessary, the determination of the weight of powder contained in a cylindrical graduate of known volume, or a standard pint or quart measure, may be made. In all cases the gravimetric density is the ratio expressed by dividing the weight of powder required to fill the measure even full by the weight of water from the same measure even full.

The gravimetric density of black powders varies over a considerable range, from about 1 to 1.3.

ABSOLUTE DENSITY.

By "absolute density" is meant the true specific gravity of the powder, the air space between grains being disregarded and only the density of the grains being considered. Owing to the fact that both sodium and potassium nitrates are readily soluble in water, it is not possible to make this determination with black powder by the picnometer method with water, so that a number of instruments involving the use of mercury have been designed by different investigators.

The form of instrument used by the Bureau of Mines is illustrated in figure 5, and was devised by one of the authors.[a] The apparatus consists of a reservoir and means for introducing within this reservoir a picnometer bottle containing the powder whose density is to be determined. By opening the water-supply valve the mercury is raised within the reservoir, and on closing it and opening the waste-pipe valve the water pressure is removed from the mercury in the lower reservoir. A Torricellian vacuum is thus created in the dome above the bottle; most of the air in the bottle escapes and is replaced with mercury. The operation is repeated until the air in the bottle is completely replaced. With this instrument the determination of the absolute density of black powder may be quickly and accurately made.

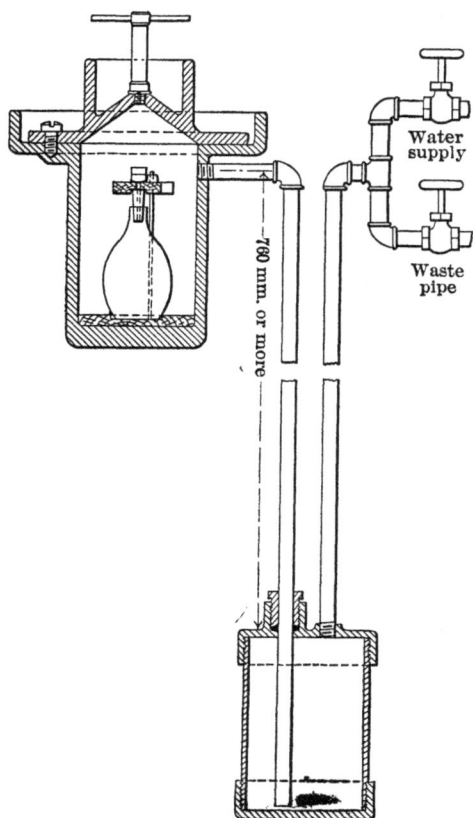

FIGURE 5.—Densimeter.

SAMPLING.

About 50 to 100 grams of the original sample is crushed in small portions in a porcelain mortar and passed through an 80-mesh sieve. All precautions are taken to avoid unnecessary exposure of the sample to the air during this treatment. If each portion is placed in a stoppered bottle as soon as sifted, there is no appreciable change in hygroscopic moisture content. The powdered sample is well mixed before its analysis is begun.

---

[a] Snelling, W. O., Improved densimeter. Proc. 8th Int. Cong. App. Chem., vol. 4, 1912, p. 105. (Chem. Abs., vol. 6, 1912, p. 3524.)

CHEMICAL EXAMINATION.

The chemical examination of black powder consists essentially of the determination of the amounts of moisture, nitrate, sulphur, and charcoal present. The nitrate is readily separated from the sulphur and charcoal by the solvent action of water, and after the residue from water extraction has been carefully dried the sulphur may be readily separated from the charcoal by the action of carbon bisulphide, or other solvent.

The charcoal is always determined by difference. After being dried the charcoal is usually ignited and its content of ash determined.

DETERMINATION OF MOISTURE.

The determination of moisture is carried out exactly as has been described in the analysis of dynamite (p. 20), a 2-gram sample being spread on a 3-inch watch glass and desiccated for three days over sulphuric acid. It is customary in some explosives laboratories to determine moisture on a sample that is crushed only sufficiently to pass through a 10 to 12 mesh sieve, because in further pulverization the moisture content of the powder may be influenced by atmospheric conditions. Comparative determinations have indicated, however, that unless there is undue exposure in preparing the sample the difference in moisture content between the coarse and the finely powdered sample is slight, and since the nature of black powder is such that a finely powdered sample must be used for chemical analysis it is considered much more convenient to use the same for the moisture determination.

Many authorities recommend that moisture in black powder be determined by drying in an oven at temperatures of 60° to 100° C.[a] As sulphur is more or less volatile at temperatures even slightly above ordinary, the authors thought it advisable to compare the relative merits of oven drying with the desiccation method. A series of determinations was therefore made on large, well-mixed samples of finely ground powder (80-mesh), 2 grams spread uniformly in a thin layer on a 3-inch watch glass being used for each determination. The samples heated in ovens were allowed to cool for 15 minutes in desiccators before weighing. It was found necessary to make the weighings as rapidly as possible in order to prevent increase of weight. The following results were obtained.

[a] Lunge, G., and Berl, E., Chemisch-technische Untersuchungsmethoden, vol. 3, 1910, p. 116; Guttmann, O., Schiess- und Sprengmittel, 1900, p. 48.

Results of determinations of moisture in black powder.

SAMPLE A.

[Determinations by W. C. Cope.]

Time.	Loss on desiccation over sulphuric acid.	Loss on drying at 70°.	Loss on drying at 100°.
Hours.	*Per cent.*	*Per cent.*	*Per cent.*
1	....	1.00	1.05
2	....	1.00	1.10
3	....	1.00	1.15
5	....	1.00	1.30
72	1.00	....	....

SAMPLE B.

[Determinations by C. A. Taylor.]

1	....	0.44	0.57
2	....	.46	.78
3	....	.50	.94
5	....	.54	1.07
7	....	.59	1.19
24	0.42—0.47	.74	3.70
72	0.46—0.49	....	....

The analysis (calculated moisture free) of sample B was originally as follows: Sodium nitrate, 74.07; sulphur, 10.09; charcoal 15.84. The sample dried 24 hours at 100°, with a loss of 3.70 per cent, was analyzed with the following result: Sodium nitrate, 77.0; sulphur, 6.53, charcoal 16.47. It is therefore evident that there is a loss of sulphur from black powder at 100°, and that this loss is appreciable in even a few hours' heating, whereas at 70° the loss for periods of heating up to five hours is approximately the same as the moisture determined by desiccation.

Desiccation gives practically constant weight in 24 hours, but as the loss of moisture takes place more slowly in coarser samples, a uniform period of three days has been adopted.

Sulphur alone is slightly affected by temperatures up to 100° C. A sample of approximately 5 grams of powdered brimstone (80-mesh) was desiccated for two days over sulphuric acid without loss of weight. It was then dried five hours at 70° C., losing only 0.003 per cent; further drying for five hours at 97° caused a loss of only 0.01 per cent.

EXTRACTION WITH WATER; DETERMINATION OF NITRATES.

In the determination of nitrates by extraction with water, about 10 grams of the finely ground sample is weighed in a Gooch crucible with asbestos mat and about 200 c. c. of water, in successive portions of 15 to 20 c. c. each, is drawn through the sample by means of suction. The complete solution of the nitrate is hastened by the use of warm or hot water, although 200 c. c. of cold water is usually sufficient. The final portions of water passing through the crucible should be tested for soluble nitrate by evaporation on a glass plate, or an excess of strong sulphuric acid containing a few crystals of diphenylamine may

be added to a few drops of the water, and an intense blue coloration will indicate the presence of nitrate.

The extraction is made on duplicate samples as with dynamite. After the complete removal of the nitrate the crucibles containing the portion insoluble in water are placed in a drying oven at a temperature of about 70° and dried to constant weight, usually overnight, although five hours is generally sufficient. The percentage of loss of weight, minus the moisture content found as described above, represents the total water-soluble material, and includes, in addition to sodium or potassium nitrate, a small amount of water-soluble organic material from the charcoal and the impurities in the original nitrate, such as chlorides and sulphate. An aliquot portion of the water extract is evaporated to dryness on a steam bath, treated with a little nitric acid, again evaporated, heated to slight fusion, and weighed. (See p. 44.) For accurate analysis the amounts of chlorides and sulphates may be determined in separate portions of the water extract and the true nitrate content determined by difference, or a direct determination of nitrate may be made with the nitrometer on a portion of the extract, as previously described.

The water solution should, of course be tested to determine whether sodium or potassium nitrate is present. This determination is conveniently made by heating to redness a clean platinum wire dipped in the solution, and observing the color of the flame through several thicknesses of cobalt glass. Potassium is indicated by its characteristic red color, and the yellow of the sodium flame is entirely cut off by the blue glass. Without the cobalt glass a sodium nitrate powder should give an intense yellow flame and a potassium nitrate powder a pale pink or lavender flame. If both sodium and potassium nitrates are indicated, a determination of potassium is best made by the sodium-cobalti-nitrate method of Drushel,[a] or the proportions of sodium and potassium nitrates may be calculated with an approximate degree of accuracy from the total weight of nitrates found by evaporation and the percentage of nitrogen in the combined weight of these nitrates as determined by the nitrometer.

The following illustrates the method of calculation employed:

Let a = weight of both nitrates.

x = weight of sodium nitrate.

Then $a - x$ = weight of potassium nitrate.

Let b = percentage of nitrogen found in combined nitrates.

16.47 = percentage of nitrogen in sodium nitrate.

13.87 = percentage of nitrogen in potassium nitrate.

$$\text{Then } 0.1647x + 0.1387 \ (a-x) = \frac{ba}{100.}$$

---

[a] Bowser, L. T., The determination of potassium by the cobalti-nitrate method. Jour. Ind. and Eng. Chem., vol. 1, 1909, p 791.

Solving for x gives the weight of sodium nitrate present in the mixture, and subtracting this from the total weight of the mixture gives the potassium nitrate.

EXTRACTION WITH CARBON DISULPHIDE; DETERMINATION OF SULPHUR.

The dried and weighed material left from the extraction with water consists of the sulphur and charcoal. The sulphur is determined by loss of weight on extraction with carbon disulphide in the Wiley extraction apparatus, the method being exactly the same as that used in the extraction with ether.

Because of the fact that it is difficult to obtain carbon disulphide that does not leave a residue of sulphur on evaporation, it is not customary to evaporate the carbon-disulphide extract to dryness and weigh the sulphur dissolved from the powder. This may be done, however, if freshly distilled pure carbon disulphide is used.

INSOLUBLE RESIDUE, CHARCOAL.

The residue remaining in the crucibles is weighed directly as charcoal after drying to constant weight at about 100°.

In drying the crucibles after the carbon disulphide extraction extreme care should be used to avoid setting fire to the inflammable vapor of the carbon disulphide, as it sometimes happens that the heavy vapor from the crucibles passes down to the flame by which the water oven is heated. Carbon disulphide has the lowest ignition temperature of any material not containing phosphorus, therefore in extracting with carbon disulphide, or in handling the crucibles after extraction, considerable care should be taken to avoid proximity to lights or fire.

DETERMINATION OF ASH.

The ash in the charcoal is determined by ignition over a Bunsen burner until all of the carbon has been burned off, and weighing. The ash is usually found to be about 0.5 to 1 per cent of the total powder. In case of an abnormally high ash value it is possible that the extraction with water was incomplete, leaving some nitrate undissolved.

The sulphur used in black powder is almost invariably brimstone, flowers of sulphur not being suitable for this purpose because of the invariable presence of acidity.

In those cases where flowers of sulphur are used, however, it should be noted that the extraction of the sulphur with carbon disulphide will be incomplete. Watts,[a] Gody,[b] and other authorities have called attention to the fact that flowers of sulphur always contain a con-

a Watts's Dictionary of Chemistry, 1905, vol. 4, pp. 606-610.

b Gody, L., Traité théorique et pratique des matières explosives, 1907, p. 99.

siderable amount of insoluble amorphous sulphur, often amounting to as much as 35 per cent.

Experiments made in the bureau's explosives laboratory with samples of brimstone and flowers of sulphur gave results as follows:

Relative solubility of flowers of sulphur and brimstone in carbon disulphide.

	Weight of sample.	Loss of weight on extraction.[a]	Insoluble substances.
	Grams.	*Grams.*	*Per cent.*
Flowers of sulphur...	1.0907 1.0450	0.7817 .7472	23.21 28.50
Brimstone...	1.1238 1.1004	1.1230 1.0996	.07 .07

[a] Extracted two hours in Wiley extraction apparatus with carbon disulphide.

In view of the fact that when flowers of sulphur are present the extraction with carbon disulphide is incomplete, some authors have recommended the use of hot aniline as a solvent for the sulphur.

The solubility of sulphur in various solvents is shown in the following table: [a]

Solubility of sulphur in 100 parts (by weight) of various solvents.

Solvent.	Temperature.	Parts (by weight) of sulphur dissolved.
	° C.	
Carbon disulphide...	0	23.99
Do...	15	41.65
Do...	22	46.05
Do...	38	94.57
Do...	[a] 55	181.34
Benzene...	26	.965
Do...	71	4.377
Toluene...	23	1.479
Ether...	23.5	.972
Chloroform..	22	1.205
Phenol..	174	16.35
Aniline...	130	85.96

[a] Boiling.

Of these solvents carbon disulphide and aniline are the ones that would appear to be of greatest practical value in analysis. Carbon tetrachloride, chloroform, and benzene have also been used with success as solvents for sulphur, but their use requires long-continued extraction.

Experiments have been made in the explosives laboratory of the bureau to determine the suitability of aniline as a solvent for sulphur in the analysis of black powder.

Extractions of both flowers of sulphur and brimstone with aniline heated to 130° to 140° showed only about 0.05 of 1 per cent insoluble material in each.

[a] Gody, L., Traité théorique et pratique des matières explosives, 1907, p. 85. Biedermann, R., Chemiker Kalender, pt. 1, 1910, p. 291.

Several samples of black powder were analyzed, the sulphur being determined by means of extraction with hot aniline, and the results were compared with those obtained by the usual method of extraction with carbon disulphide. The extraction with the aniline was made by adding 10-c. c. portions of aniline, previously heated to 130° to 135° C. to the crucibles containing samples that had been extracted with water and dried in the usual manner, the hot aniline being drawn through by suction; this treatment was repeated from 5 to 12 times, using a total of 50 to 125 c. c. of aniline in different determinations. The last portions of aniline were removed by washing with a small amount of alcohol, and the residue dried at 100° C. The loss of weight was considered as the amount of sulphur present.

Comparative determinations of sulphur in black powder by extraction with carbon disulphide and aniline.

Sample No.	Sulphur extracted.	
	With carbon disulphide.	With aniline (130°).
1	13.64	13.79
2	9.99	9.72
3	7.81	7.74

Each of the above figures is the average of four to six closely agreeing results. This comparison shows that the aniline method gives very satisfactory results. Even as small an amount as 50 c. c. of aniline was found to be sufficient for complete extraction if each portion was allowed to stand a short time before sucking dry. The aniline may be readily recovered by distillation and used repeatedly.

A series of experiments were made to determine whether any difference in results would be effected by extracting the sulphur with carbon disulphide before the nitrate was extracted with water—that is, an inversion of the order of the extractions. A number of samples of black powder were analyzed by both methods with the results noted in the following tables. For comparison, determinations were made of the sulphur by precipitation as barium sulphate after oxidation with nitric acid and potassium chlorate.

Results of analyses of black powder.

[Samples A, B, and C analyzed by C. A. Lambert; sample D analyzed by C. A. Taylor.]

METHOD 1 (WATER EXTRACTION FIRST).

Sample.	Moisture.	Water extract.	Carbon disulphide extract.	Insoluble residue.	Sulphur determined as $BaSO_4$.
	Per cent.	*Per cent.*	*Per cent.*	*Per cent.*	*Per cent.*
A	0.16	69.76	13.64	16.44	13.55
	.16	69.73	13.55	16.56	
B	.21	74.07	10.09	15.63	10.05
	.21	74.27	9.89	15.63	
	.36	78.88	7.87	12.89	
C	.36	78.98	7.68	12.98	7.89
	.36	78.91	7.89	12.84	
	.47	73.84	10.02	15.67	
D	.47	73.74	10.06	15.73	....
	.47	73.79	10.04	15.70	

METHOD 2 (CARBON DISULPHIDE EXTRACTION FIRST).

A	0.16	69.58	13.74	16.52	13.55
	.16	69.57	13.80	16.47	
B	.21	73.97	10.39	15.43	10.05
	.21	73.92	10.44	15.43	
C	.36	78.45	8.47	12.72	7.89
	.36	78.54	8.49	12.61	
	.47	73.91	10.15	15.47	
D	.47	73.74	10.25	15.54	....
	.47	73.82	10.20	15.51	

It will be noted that in every case a greater loss was obtained on extracting with carbon disulphide when such extraction preceded the extraction with water than when it followed the extraction with water. That a more correct value for the amount of sulphur present is obtained by first removing the water-soluble portion of the powder, is shown by the results of the gravimetric determination of sulphur as barium sulphate. The latter results agree closely with the loss on extraction with carbon disulphide after removal of the nitrate.

The differences in results noted above are not readily explained. That both sodium and potassium nitrates are practically insoluble in carbon disulphide was shown by extracting a number of dried samples of each of these nitrates with this solvent, losses of only 0.01 to 0.05 per cent being obtained. It was further demonstrated that the small amount of moisture present in the powder samples at the time of the extraction with carbon disulphide was not the cause of the high results of method 2. This was shown by extracting several samples of black powder with carbon disulphide both before and after drying in vacuum desiccators. On correcting all results to the sample in original condition it was found that the loss on extraction was practically the same in both cases. The results are shown in the table following.

Effect of moisture in black powder on extraction with carbon disulphide before removal of nitrate.

[Determinations by C. A. Taylor.]

Sample.	Moisture.		Loss on extraction with carbon disulphide.	
	Original sample.	Desiccated sample.	Original sample.	Desiccated sample.
	Per cent.	*Per cent.*	*Per cent.*	*Per cent.*
E	0.30	....	13.82	13.93
F	.22	....	10.02	9.99
G	.32	....	7.94	7.92

The method adopted by the Bureau of Mines for the analysis of black powder is as follows:

BUREAU OF MINES METHOD OF ANALYSIS.

Moisture is determined by desiccating a 2-gram portion of the 80-mesh sample spread uniformly on a 3-inch watch glass for three days in a sulphuric-acid desiccator. Nitrates are determined by extraction with water, sulphur by extraction with carbon disulphide, and charcoal by weighing the dried insoluble residue. The soluble impurities (sulphates, chlorides, etc.) are determined separately and a direct determination of nitrate made by means of the nitrometer.

In the analysis of black powder, as in the analysis of all other explosives, the Bureau of Mines follows the standard methods outlined in this bulletin except when the presence of unusual constituents renders necessary additional separations or determinations or introduces new difficulties. As these conditions are seldom met in the analysis of the more simple explosives, such as dynamite and gelatin dynamite, but are not infrequently found in connection with short-flame explosives intended for use in coal mining, they will be taken up in a subsequent bulletin, which will consider the analysis of permissible explosives.

PUBLICATIONS ON MINE ACCIDENTS AND TESTS OF EXPLOSIVES.

The following Bureau of Mines publications may be obtained free by applying to the Director, Bureau of Mines, Washington, D. C.:

BULLETIN 10. The use of permissible explosives, by J. J. Rutledge and Clarence Hall. 1912. 34 pp., 5 pls.

BULLETIN 15. Investigations of explosives used in coal mines, by Clarence Hall, W. O. Snelling, and S. P. Howell, with a chapter on the natural gas used at Pittsburgh, by G. A. Burrell, and an introduction by C. E. Munroe. 1911. 197 pp., 7 pls.

BULLETIN 17. A primer on explosives for coal miners, by C. E. Munroe and Clarence Hall, 61 pp., 10 pls. Reprint of United States Geological Survey Bulletin 423.

BULLETIN 20. The explosibility of coal dust, by G. S. Rice, with chapters by J. C.W. Frazer, Axel Larsen, Frank Haas, and Carl Scholz. 204 pp., 14 pls. Reprint of United States Geological Survey Bulletin 425.

BULLETIN 44. First national mine-safety demonstration, Pittsburgh, Pa., October 30 and 31, 1911, by H. M. Wilson and A. H. Fay, with a chapter on the explosion at the experimental mine, by G. S. Rice. 1912. 75 pp., 7 pls.

BULLETIN 46. An investigation of explosion-proof mine motors, by H. H. Clark. 1912. 44 pp., 6 pls.

BULLETIN 48. The selection of explosives used in engineering and mining operations, by Clarence Hall and S. P. Howell. 1913. 50 pp., 3 pls.

TECHNICAL PAPER 4. The electrical section of the Bureau of Mines, its purpose and equipment, by H. H. Clark. 1911. 12 pp.

TECHNICAL PAPER 6. The rate of burning of fuse as influenced by temperature and pressure, by W. O. Snelling and W. C. Cope. 1912. 28 pp.

TECHNICAL PAPER 7. Investigations of fuse and miners' squibs, by Clarence Hall and S. P. Howell. 1912. 19 pp.

TECHNICAL PAPER 11. The use of mice and birds for detecting carbon monoxide after mine fires and explosions, by G. A. Burrell. 1912. 15 pp.

TECHNICAL PAPER 12. The behavior of nitroglycerin when heated, by W. O. Snelling and C. G. Storm. 1912. 14 pp., 1 pl.

TECHNICAL PAPER 13. Gas analysis as an aid in fighting mine fires, by G. A. Burrell and F. M. Seibert. 1912. 16 pp.

TECHNICAL PAPER 17. The effect of stemming on the efficiency of explosives, by W. O. Snelling and Clarence Hall. 1912. 20 pp.

TECHNICAL PAPER 18. Magazines and thaw houses for explosives, by Clarence Hall and S. P. Howell. 1912. 34 pp., 1 pl.

TECHNICAL PAPER 19. The factor of safety in mine electrical installations, by H. H. Clark. 1912. 14 pp.

TECHNICAL PAPER 21. The prevention of mine explosions; report and recommendations, by Victor Watteyne, Carl Meissner, and Arthur Desborough. 12 pp. Reprint of United States Geological Survey Bulletin 369.

TECHNICAL PAPER 23. Ignition of mine gas by miniature electric lamps, by H. H. Clark. 1912. 5 pp.

TECHNICAL PAPER 24. Mine fires; a preliminary study, by G. S. Rice. 1912. 51 pp.

TECHNICAL PAPER 28. Ignition of mine gas by standard incandescent lamps, by H. H. Clark. 1912. 6 pp.

TECHNICAL PAPER 29. Training with mine-rescue breathing apparatus, by J. W. Paul. 1912. 16 pp.

MINERS' CIRCULAR 3. Coal-dust explosions, by G. S. Rice. 1911. 22 pp.

MINERS' CIRCULAR 4. The use and care of mine-rescue breathing apparatus, by J. W. Paul. 1911. 24 pp.

MINERS' CIRCULAR 5. Electrical accidents in mines; their causes and prevention, by H. H. Clark, W. D. Roberts, L. C. Ilsley, and H. F. Randolph. 1911. 10 pp., 3 pls.

MINERS' CIRCULAR 6. Permissible explosives tested prior to January 1, 1912, and precautions to be taken in their use, by Clarence Hall. 1912. 20 pp.

MINERS' CIRCULAR 9. Accidents from falls of roof and coal, by G. S. Rice. 1912. 16 pp.

MINERS' CIRCULAR 10. Mine fires and how to fight them, by J. W. Paul. 1912. 14 pp.

MINERS' CIRCULAR 11. Accidents from mine cars and locomotives, by L. M. Jones. 1912. 16 pp.

INDEX.

79

ANALYSIS OF EXPLOSIVES[1]

C. G. STORM[2]

The methods described in this chapter cover all of the more common types of explosives employed in the United States for both commercial and military purposes. Those of chief commercial importance are black powder, nitroglycerin dynamites, including "straight" dynamites, ammonia dynamites, gelatin dynamites, and low-freezing dynamites, "Permissible" coal mining explosives and nitrostarch blasting explosives. Military explosives include smokeless powder, guncotton, trinitrotoluene, picric acid, ammonium picrate, "Amatol," tetryl and tetranitroaniline. No sharp distinction can, however, be drawn between commercial and military explosives, as many are utilized for both purposes. Many tons of surplus TNT. have been used recently in commercial work; nitrostarch explosives found important application for military use during the war; mercury fulminate and other detonators and priming compositions are essential in every field of explosives.

The methods described have largely been used by the writer in practical explosives testing and analysis in connection with both Government and private work. Most of those applying to commercial explosives have been approved by the United States Bureau of Mines for use in its Explosives Chemical Laboratory.[3]

BLACK POWDER

The composition of black powder varies to some extent, depending chiefly on the purpose for which the explosive is to be used. Black blasting powder contains sodium nitrate, charcoal and sulphur; black gunpowder is quite similar except that potassium nitrate is generally substituted for the sodium nitrate; black fuse powder is similar to the latter, differing mainly in its granulation. The same general method of analysis is therefore applicable to all types of black powder.

Sampling. From 50 to 100 grams of the original sample is crushed in small portions in a porcelain mortar and completely passed through an 80-mesh sieve, care being taken to avoid undue exposure to the air. The separate powdered portions are promptly bottled and the entire sample is finally well mixed.

Moisture. The standard method of the Bureau of Mines is to desiccate a 2-gram sample on a 3-inch watch glass over sulphuric acid for three days, the loss of weight being moisture. It has been shown, however, that equally accurate results can be obtained by drying at 70° C. in a constant temperature oven to constant weight, for which 2–3 hours is usually sufficient. As much as 5 hours drying at 70° C. will not cause loss of sulphur. Drying at 100° C. gives results which are slightly high, due to loss of sulphur.

[1] Received March, 1920. Published by permission of Chief of Ordnance, U. S. A.

[2] Professor of Chemical Engineering, Ordnance School of Application, Aberdeen Proving Ground, Maryland.

[3] See Bureau of Mines Bulletin No. 51, "The Analysis of Black Powder and Dynamite," W. O. Snelling and C. G. Storm, 1913, and Bulletin No. 96, "The Analysis of Permissible Explosives," C. G. Storm, 1916.

Nitrates. About 10 grams of the finely ground sample in a Gooch crucible provided with an asbestos mat, is extracted with warm water by means of suction, the water being added in 15–20 cc. portions and each portion being allowed to stand in the crucible a short time before suction is applied. About 200 cc. of water is usually sufficient, but the last drops of filtrate should be tested by evaporation to ensure the absence of nitrates. A blue color on the addition of sulphuric acid containing a few crystals of diphenylamine will also indicate the presence of nitrates.

The water extract includes a small amount of water-soluble organic material from the charcoal in addition to the nitrate. It is made up to 250 cc. and an aliquot portion (50 cc.), evaporated to dryness on the steam bath, treated with a little nitric acid, again evaporated, heated to slight fusion and weighed.

If allowance for impurities in the nitrate is desired, a direct determination of nitrate may be made on a separate portion of the water extract by the Devarda method or by means of the nitrometer, but for all practical purposes the evaporation method is sufficient. The usual tests should be made to determine whether sodium nitrate or potassium nitrate is present.

The residue left in the crucible, consisting of sulphur and charcoal, is dried at about 70° C. to constant weight (for 5 hours or over night if more convenient), the loss of weight minus the moisture content being the water-soluble portion. This result serves as a check on the evaporation result.

Sulphur. The residue in the crucible is extracted in a Wiley extractor or other continuous extraction apparatus with carbon disulphide, until evaporation of a small portion of the solvent passing through the crucible shows absence of sulphur. The excess of carbon disulphide is then allowed to evaporate from the crucible in a warm place away from flame, and the residue finally dried to constant weight at 100° C. The loss of weight is considered as sulphur.

Charcoal. The dry residue in the crucible should consist only of charcoal.

Ash. The ash in the charcoal may be determined by ignition over a Bunsen burner until all of the carbon has been burned off, and weighing. This ash also contains, of course, any non-volatile matter that may have been present in the sulphur and nitrate.

Calculation of Results. Since a portion of the charcoal is always dissolved in the water extract, it is customary to express the content of charcoal by subtracting the sum of the following from 100%:

% Moisture (by desiccation or drying at 70° C.).
% Nitrate (by evaporation of water extract with HNO_3).
% Sulphur (by loss on extraction with CS_2).

NITROGLYCERIN DYNAMITES

"Straight" Dynamite

So-called "Straight" nitroglycerin dynamite has been manufactured to only a relatively small extent in this country during the past few years, owing to the high cost of glycerin. It has been largely replaced by the ammonia and low-freezing dynamites, in which a large part of the nitroglycerin is replaced by ammonium nitrate and nitrosubstitution compounds. Furthermore, developments in the manufacture of both ammonium nitrate and nitrocompounds during the war have rendered unlikely any great increase in the manufacture of straight dynamites. They are still largely used, however, where quick-acting blasting explosives of high strength are required, as in work in hard rock. They consist essentially of nitroglycerin absorbed in a "dope" composed of a combustible absorbent, usually wood pulp, and an oxidizing material (sodium nitrate), to which is added a small amount of an antacid (calcium carbonate, zinc oxide, etc.). The analysis is best carried out by successive extractions, usually with ether, water, and dilute hydrochloric acid.

Sampling. The wrappers are removed from a number of the cartridges, and from 3 to 5 cm. of the ends of the exposed roll of explosives rejected. The remainder is thoroughly mixed on a large sheet of paraffined paper or in a large porcelain dish, and an average sample selected and bottled—usually about one half pound. The importance of thorough mixing of the sample must not be overlooked, in view of the fact that there is frequently a decided tendency for the nitroglycerin to segregate due to insufficient or unsuitable absorbent, so that this liquid ingredient may not be uniformly distributed throughout the cartridge. Also if a carefully mixed sample has been allowed to stand for some days, especially in a warm place, segregation may occur in the bottle, so that it is advisable to again mix the sample before analysis.

Qualitative Examination. Although a qualitative analysis of a sample known to be straight nitroglycerin dynamite is usually unnecessary, the exact nature of the sample may be unknown, and a knowledge of the composition of some of the more complex types of dynamite is necessary before a quantitative analysis can be properly conducted.

About 25 grams of the sample is shaken with several successive portions of ether in a large stoppered test tube, the ether being decanted off through a filter paper and the residue finally washed on the filter. The ether solution is allowed to evaporate slowly on a steam bath and the filter paper spread out on a glass plate in an oven so that the residue may dry quickly. The evaporated ether extract may contain nitroglycerin, sulphur (especially in the lower grades of dynamite), rosin, vaseline, or paraffin oil (in ammonia dynamite), nitrotoluenes and other nitrocompounds (in low-freezing dynamites), etc.

Nitroglycerin is readily detected by shaking a drop of the liquid with one or two cc. of strong H_2SO_4 and about 1 cc. of mercury in a test tube, an evolution of brown fumes of nitric oxides being noted if nitroglycerin is present. Sulphur will appear as crystals in the evaporated extract, and may be identified by removing them, washing with acetic acid, and noting the odor of SO_2 on heating in a flame. Rosin, vaseline, oils, etc., appear as a greasy scum on the surface of the nitroglycerin or adhering to the walls of the beaker. These substances, like sulphur, are practically insoluble in acetic acid (70%), and may be separated from the nitroglycerin by means of this solvent. Trinitrotoluene will appear in the nitroglycerin as long yellowish needles, which may be removed, recrystallized from alcohol, and identified by their melting point (approx. 80° C.), or by the red color produced when the alcoholic solution is treated with a little caustic soda solution.

The residue insoluble in ether is replaced in the test tube and treated with water in a similar manner until all water-soluble material has been dissolved. The water solution is tested for sodium, potassium, barium, zinc, etc., and for nitrates, chlorides, etc., using the general methods of qualitative analysis.

The residue is again treated with cold dilute HCl, any effervescence being noted as indicating the presence of a carbonate, and the resulting solution tested for calcium, magnesium, zinc, etc., which may have been present as carbonates or oxides for the purpose of serving as antacids.

The residue insoluble in ether, water, and cold acid may contain wood pulp, starchy cereal products, sawdust, nitrocellulose, ground vegetable ivory (button waste), kieselguhr, ground nut shells, etc. It is most conveniently examined by means of a low-power microscope, whereby its constituents are usually readily determined. Starch is easily detected by heating a portion to boiling with dilute acid, cooling and adding a few drops of iodine solution (in KI), a blue coloration indicating starch.

Moisture. Moisture is best determined by desiccation over sulphuric acid, a sample of about 2 grams being spread evenly over the surface of a 3-inch watch glass and desiccated for 3 days. Continued desiccation causes a gradual loss of nitroglycerin, but the 3-day loss may be safely assumed to closely represent the actual moisture content. The time of the determination may be greatly shortened by the use of a vacuum desiccator, in which case 24 hours desiccation will give a close approximation to the true moisture content.

It must be remembered that in determining moisture in the presence of nitroglycerin, some volatilization of the latter is unavoidable, and that therefore the method followed must be an empirical one. An attempt to desiccate the sample to constant weight will show that there is undoubtedly a continual loss of nitroglycerin. This has been demonstrated[1] by a series of weighings of a sample exposed for a period of 459 days at a constant temperature of 33°–35° C. in an empty desiccator containing no desiccating agent. A gradual loss resulted during the entire period, totaling 17.52% of the original weight of the sample, the original moisture content of which was about 1%.

Extraction with Ether. Ether removes from dynamite not only the nitroglycerin, but, as has already been mentioned, sulphur, resins (present as a component or as a constituent of the wood pulp), oils (usually from cereal products present), etc. Nitrotoluenes, paraffin, vaseline, etc., are not normal constituents of straight dynamite and are considered under the type of explosive in which they are most likely to occur.

Reflux Condenser Method. From 6 to 10 grams of the sample is weighed in either a porcelain Gooch crucible with asbestos mat or a porous alundum filtering crucible of about 25 cc. capacity. The asbestos mat is best prepared as follows: A mixture of 1 liter of water and 5 grams of previously ignited and shredded short fibre asbestos free from hard lumps and very fine material is well shaken and about 10 cc. poured into the crucible. Suction is applied

[1] Storm, C. G., "The Analysis of Permissible Explosives," Bulletin No. 96, Bureau of Mines, pages 21–24, 1916.

and a smooth and perfect mat almost invariably results. The crucibles thus prepared are dried at 100° and are ready for use.

The sample in the extraction crucible is extracted with about 35 cc. of ether (U. S. P.) preferably in a continuous extraction apparatus (Wiley or similar type preferred), for about 45 minutes to 1 hour, water being continuously circulated through the condenser and the extraction tube heated on a water bath, or electric heater, the temperature of which is so regulated that the sample in the crucible will be kept covered with ether without overflowing.

Suction Method. If desired, the ether extraction may be carried out by suction, the Gooch crucible being held in a carbon tube passing through the stopper in a suction flask. About 100 cc. of ether in 6 to 8 portions is passed through the crucible, each portion being allowed to stand in the crucible for one minute before applying gentle suction. No more air than is necessary should be drawn through the sample in order to avoid condensation of moisture in the sample, which might dissolve a portion of the water-soluble salts. This method uses considerably more ether than the reflux condenser method and its chief advantage is that the apparatus required is more simple.

On completion of the extraction the crucible is at once placed in a drying oven, or the excess ether may be removed by suction before drying. If ammonium nitrate is present the drying should be conducted at 70° C. for 18 hours or overnight, but otherwise 5 hours at 100° C. is ample. The loss of weight represents all ether-soluble material plus the moisture in the original sample.

Evaporation of Ether Extract. The ether extract is washed out of the extraction tube or suction flask with a little ether into a tared evaporating dish or small beaker and the ether allowed to evaporate spontaneously in a warm place, or evaporated by means of the "bell jar evaporator."[1] The latter consists of a tubulated bell jar with openings at top and side, placed on a ground glass plate, a slow current of dry compressed air from two drying cylinders containing H_2SO_4 and soda lime respectively, entering the top opening through a glass tube, the lower end of which extends to about one half inch from the surface of the ether solution in the beaker, which is placed on the glass plate. The dry air current striking the surface of the solution with just enough force to cause a slight "dimple," causes rapid evaporation of the ether, and deposition of moisture in the beaker along with the evaporated residue is avoided. The low temperature produced by the rapid evaporation minimizes the loss of nitroglycerin by volatilization. From 5 to 6 hours is usually required for complete evaporation, which should be determined by check weighings. If the bell jar method is not used, the residue, after removal of the ether, must be desiccated over H_2SO_4 for at least 24 hours in order to remove moisture deposited during evaporation.

Nitroglycerin. Nitroglycerin is determined in the dried and weighed ether extract from which all ether has been removed as above described. This determination is best made by means of the du Pont modification of the 5-part Lunge nitrometer (see p. 354, Vol. I). The sample is dissolved in 5–10 cc. of pure sulphuric acid (specific gravity 1.84) and transferred to the generating bulb of the nitrometer, the beaker and cup of the nitrometer being washed with several further additions of acid until a total of 20–25 cc. has been used.

[1] Storm, C. G., "The Analysis of Permissible Explosives," Bulletin, 96, Bureau of Mines, page 35, 1916.

If the quantity of nitroglycerin present is too great, the sample, dissolved in sulphuric acid, is transferred to a burette and an aliquot part run into the nitrometer. The maximum amount of pure nitroglycerin used for the determination should not exceed 0.75 gram. The determination is carried out in the usual manner and the reading of the gas volume in the graduated reading tube divided by .1850 to find the weight of nitroglycerin in the sample used for the determination (pure nitroglycerin contains 18.50% N).

Sulphur, Resins, Oils, etc. It is always preferable to carry out the extraction with ether on duplicate samples, using one sample of the extract for the determination of nitroglycerin as above, and the other for determining sulphur, resins, oils, etc., that may also be contained in the ether extract.

The weighed extract is redissolved in a mixture of ether and alcohol, previously neutralized with standard alkali. The solution thus obtained is titrated with standard alcoholic potash solution using phenolphthalein indicator. 1 cc. of tenth normal alkali is equal to 0.034 grams of rosin (colophony).

A large excess of the alcoholic potash is now added and the mixture heated several hours or overnight on the steam bath to saponify the nitroglycerin. Shake with water and ether in a separatory funnel. The ether solution contains paraffin, vaseline, or mineral oils that may be present, and is evaporated and the residue weighed. The water solution is acidified with HCl, and Br added to oxidize any sulphur. Any separated rosin is filtered off and weighed as a check on the titration, and sulphur determined in the filtrate by precipitation as $BaSO_4$.

Sulphur may also be separated from nitroglycerin by means of acetic acid of approximately 70% strength, the nitroglycerin being quite soluble in acetic acid and the sulphur almost insoluble. The sulphur is filtered from the solution, washed slightly with alcohol to remove the acetic acid solution, dried and weighed.

If a considerable quantity of crystals of sulphur is found in the evaporated ether extract, it is possible that all of the sulphur has not been removed by the ether, and in this case an extraction is made with carbon disulphide, in exactly the same manner as the ether extraction. This extraction is made subsequent to the extraction with water, the sulphur being determined by loss of weight of the residue or by direct weight after evaporation of the carbon disulphide away from free flame.

Extraction with Water and Determination of Nitrates. The dried and weighed residue left in the crucible after extraction with ether, is extracted with water, using a suction flask fitted with a carbon filter tube in which the crucible is held by a short length of thin-walled rubber tubing. Cold water is used for this extraction, as hot water would gelatinize any starch present. A total of at least 200 cc. of water is passed through the sample, in at least 10 portions, each portion being allowed to stand in contact with the residue for a few minutes before being sucked into the flask. An evaporation test of a few drops of the filtrate will determine the completeness of the extraction. When the extraction is complete, the crucible with its insoluble residue is dried for 5 hours, or overnight, at 95°–100° C., and the loss of weight noted as total water-soluble material. This includes nitrates and other soluble salts that may be present, together with water extract from the wood pulp, flour or other absorbent. This soluble organic material may amount to as much as 2% of the total sample, when cereal products are present. Calcium, magnesium, or zinc may also be present in solution, resulting from the action of acid decomposition products of the nitroglycerin on the carbonate or other antacid present. In routine analyses of ordinary dynamite, the loss of weight on extraction with water is usually considered as the alkaline nitrate (sodium or potassium), but where more exact results are desired an aliquot portion of the extract is evaporated to dryness with a little nitric acid to oxidize organic materials, and the residue weighed as alkaline nitrate. This weight may be corrected for inorganic impurities—chlorides, sulphates, iron, aluminum, calcium, etc.—determined separately by the usual methods.

Nitrates may be determined by means of the nitrometer, using an aliquot portion of water extract estimated to contain .6 to .8 gram of $NaNO_3$ or .8 to 1.0 gram of KNO_3. This is evaporated on the steam bath almost to dryness and transferred with as little water as possible, to the cup of the nitrometer. This solution is drawn into the generator and 30 to 40 cc. of 95–96% H_2SO_4 added slowly so as to avoid generating sufficient heat to crack the glass. The generator is then shaken for a total time of 8–10 minutes in order to be certain that the generation of gas is complete with the diluted acid. The gas is measured and the % of nitrate calculated as in the case of nitroglycerin.

Extraction with Acid. When starch is not present in the residue, a simple extraction of the residue insoluble in water is made with cold dilute HCl (1 : 10), 100 cc. being drawn through the sample in the crucible in small successive portions as described under "Extraction with Water." Several portions of water are then drawn through to wash out the acid, and the residue in the crucible dried for 5 hours at 95° to 100° C. The loss of weight is usually reported as antacid, but the base dissolved may be determined by the usual quantitative methods if desired. The acid-soluble materials generally present are calcium or magnesium carbonate or zinc oxide.

Determination of Starch. If starch is present in the residue insoluble in water, it is removed together with the antacid by boiling with dilute acid. The residue is moistened with water, scraped or washed out of the crucible into a 500 cc. beaker, the volume brought to about 250 cc. by the addition of water and 3 cc. of strong HCl, and the mixture boiled until a drop of the solution fails to give a blue color when treated on a spot plate with a drop of a solution of iodine in KI. This indicates that the starch has been completely hydrolyzed to dextrin. The mixture is then filtered through a fresh crucible, washed with water, dried and weighed, correction being made for the weight of the asbestos mat of the original crucible.

The antacid dissolved in the acid filtrate is determined as already described. The loss of weight by the boiling treatment, minus the antacid found, represents starch and other dissolved organic materials removed from cereal products or wood pulp. The insoluble residue includes the wood pulp and the crude fibre of the cereal products.

Because of the impracticability of exact separations it is customary to report all of the soluble organic material included in both water and acid extractions as "starch" or "starchy material," and the insoluble organic residue as "wood pulp and crude fibre," or the sum of these organic materials is often reported as "carbonaceous combustible material."

Insoluble Residue and Ash. The insoluble residue may contain wood pulp or sawdust, the crude fibre from various cereal products such as corn meal, wheat flour, middlings, bran, etc., ground nut shells, vegetable ivory meal, and more rarely inorganic material such as infusorial earth (kieselguhr), clay, etc. These can usually be identified by microscopic examination (see Bureau of Mines Bulletin 96, Page 74), and a determination of the ash will show whether inorganic materials are present. A high ash content may also indicate incomplete water or acid extractions.

Ammonia Dynamite

So-called ammonia dynamite is essentially "straight" dynamite in which a large part of the nitroglycerin is replaced by ammonium nitrate. The ammonium nitrate is frequently protected from moisture by a coating of vaseline or paraffin and is usually neutralized with zinc oxide. This type of dynamite generally contains less wood pulp than the corresponding grades of "straight" dynamite, and sulphur and cereal products, such as low grade flour, are usually present.

The determination of moisture and the various extractions are carried out as described for "straight" dynamite. An extraction with carbon disulphide is usually necessary to effect complete removal of the sulphur; this properly follows the extraction with water. · The analysis of the ether extract may be conducted as already described. In drying the residue left in the crucible after extraction with ether, it is important that a temperature of approximately 70° C. be used, because in the presence of ZnO, the loss of ammonium nitrate is considerable at 100° C. Pure ammonium nitrate is not appreciably affected by even 24 hours heating at 100° C., but the presence of the ZnO causes decomposition at this temperature.

The water extract contains sodium nitrate and ammonium nitrate together with practically all of the zinc oxide present, the latter ingredient being dissolved with the ammonium nitrate, and a small amount of soluble organic material from the flour or other absorbent. It is analyzed as follows: An aliquot portion is evaporated to dryness in a platinum or silica dish on a steam bath, the ammonium nitrate volatilized by careful heating over a burner, a little nitric acid added to re-oxidize any nitrate that may have been reduced to nitrite, and the residue again dried on the steam bath. The zinc oxide is now in the form of zinc nitrate and may be separated from the sodium nitrate by either of the following methods:

1. The residue is dried at 110°–120° C. and weighed as $NaNO_3$ and $Zn(NO_3)_2$. It is then dissolved in water, the zinc precipitated with Na_2CO_3, filtered, ignited and weighed as ZnO, and the $NaNO_3$ taken by difference; the total $NaNO_3$ plus $Zn(NO_3)_2$ minus $(ZnO \times 2.33) = NaNO_3$.

2. The residue is gently heated over a burner until evolution of oxides of nitrogen from decomposition of the $Zn(NO_3)_2$ has ceased, and the remaining residue weighed as $NaNO_3$ and ZnO. It is then treated with water, the insoluble ZnO filtered on a Gooch crucible, ignited and weighed, the $NaNO_3$ being taken by difference.

Ammonium nitrate is determined in a separate portion of the water extract by the usual method of distillation and titration.

The sum of the amounts of NH_4NO_3, $NaNO_3$, and ZnO found will be somewhat less than the total water extract owing to the presence of water-soluble organic material from the carbonaceous absorbents.

Gelatin Dynamite

This is a form of nitroglycerin explosive in which the nitroglycerin, instead of being absorbed in porous materials such as wood pulp, is combined with nitrocellulose in the form of a gelatinous plastic mass. As little as 3.5% of suitable grade of nitrocellulose containing about 12% nitrogen will, when heated with nitroglycerin, at about 60° C., form a jelly-like non-fluid mass when cooled to ordinary temperature. "Blasting gelatin," used to a considerable extent where great strength is required, is a stiff colloid composed of 90 to 93% nitroglycerin and 10 to 7% nitrocellulose.

All blasting explosives containing such colloids of nitroglycerin and nitrocellulose combined with an active "dope" or base, consisting of a nitrate and combustible material, are termed gelatin dynamites. This type of explosive is also known in some countries as "Gelignite."

Sampling. Owing to its pasty consistency the sample of gelatin dynamite must be prepared by cutting portions of a number of cartridges into thin bits with an aluminium or platinum spatula. The use of a steel spatula or knife for this purpose is not to be recommended for reasons of safety. An ample quantity of sample thus prepared is well mixed and bottled. Owing to its tendency to again form a solid mass upon standing, it should be analyzed as soon as possible after being prepared.

Analysis. The principal ingredients that may be found in the different types of gelatin dynamite are nitroglycerin; nitrocellulose; sulphur; rosin; sodium, potassium or ammonium nitrate; calcium or magnesium carbonate; wood pulp, cereal products and similar carbonaceous combustible materials. Low-freezing gelatins may also contain nitrotoluenes or other nitrosubstitution compounds.

Moisture is determined as described for "straight" dynamite, and the extraction with ether made in the usual manner except that ether free from alcohol (distilled over sodium) is used in order to prevent partial solution of the nitrocellulose. The latter is readily soluble in a mixture of ether and alcohol, and as ordinary U. S. P. ether contains about 4% of alcohol, there is a possibility that an appreciable part of the 0.5% to 2.0% of nitrocellulose present in the sample will be dissolved unless pure ether is used. The ether extract is evaporated and analyzed as already described and the water extraction made in the usual manner. If more than 1 or 2% of sulphur was present it will not have been completely removed by the ether, unless the extraction was continued for a sufficiently long time. In this event, it is necessary to make an additional extraction with carbon disulphide in the Wiley apparatus subsequent to extraction with water.

Nitrocellulose. After the extractions with ether, water, and CS_2 (if necessary) have been made, the nitrocellulose is determined, preferably by extraction with acetone, which is a better solvent for the purpose than a mixture of ether and alcohol. It is advisable to separate the dry residue from the crucible, leaving the asbestos mat intact if possible. The residue is transferred to a small beaker, covered with acetone and allowed to stand at least 3 or 4 hours with occasional stirring. It is then filtered through the original crucible, washed with acetone, dried and weighed, the loss of weight being regarded as nitrocellulose. To correct for small amounts of extract from the wood pulp or other carbonaceous material, the acetone solution may be evaporated to about 20–25 cc., and diluted gradually with a large volume (about 100 cc.) of hot water, which volatilizes the acetone, precipitating the nitrocellulose as a white flocculent mass, which is filtered, dried, and weighed.

The remainder of the analysis is conducted as for straight dynamite.

It will be found that the results of analysis of a gelatin dynamite do not agree with its trade markings. For example, the usual "40% strength" gelatin dynamite actually contains from 30 to 33% of nitroglycerin and about 1% of nitrocellulose. Weight for weight this explosive is considerably weaker than 40% straight dynamite, which contains 40% of nitroglycerin.

Low=Freezing Dynamite

Low-freezing dynamites vary from the dynamite types already discussed by containing an ingredient which reduces the freezing point of the nitroglycerin. This ingredient replaces a portion of the nitroglycerin which would be used in an equal grade of ordinary straight dynamite, ammonia dynamite, or gelatin dynamite. While straight nitroglycerin dynamite may freeze at temperatures as high as 8° C. (46° F.), some of the low-freezing dynamites freeze only at temperatures considerably below 0° C. Many of this type, however, cannot be relied upon to resist freezing at temperatures below the freezing point of water.

The additions made to nitroglycerin for this purpose include the nitrotoluenes, nitroxylenes, nitrohydrins, nitrosugar, and nitropolyglycerin (tetranitrodiglycern). Any of these substances present will be found in the ether extract together with, and in most cases dissolved in, the nitroglycerin.

Moisture. The determination of moisture is carried out as already described for "straight" nitroglycerin dynamite (p. 1375). Attention has been called to the fact that certain nitrosubstitution compounds, notably the mono- and dinitrotoluenes, are more or less volatile and would therefore be partly lost if the moisture is determined in a vacuum desiccator. The safest procedure is therefore to determine the moisture by desiccation for 3 days without vacuum. The difference between the total loss on extraction with ether and the direct weight of the ether extract, after evaporation of the ether in a bell-jar evaporator (p. 1376), should be equal to the moisture content of the sample. This figure will therefore serve as a check on the result obtained by desiccation.

Nitrotoluenes. Trinitrotoluene is not readily soluble in nitroglycerin and separates as crystals on evaporation of the ether from the ether extract, enabling it to be qualitatively separated and identified. It may be determined by difference, the nitroglycerin being determined by means of the nitrometer. Any dinitrotoluene present may also be determined in this manner together with the trinitrotoluene, but if mononitrotoluene is also present, the determination of the nitrogen of the nitroglycerin will be slightly in error by about 0.5530 gram of nitroglycerin for every gram of mononitrotoluene present.[1]

Mononitrotoluene is, however, seldom present except as an impurity in the so-called liquid di- and trinitrotoluenes used in low freezing dynamites, so that the determination of the nitroglycerin is usually fairly accurate and the nitrotoluenes may be calculated by difference.

[1] Storm, C. G., "The Effect of Nitrotoluenes on the Determination of Nitroglycerin by Means of the Nitrometer," Proc. 8th Int. Cong. Appl. Chem., Vol. 4, 1912, p. 117; also Bu. of Mines Bull. 41, p. 62, 1913.

The total nitrogen of the combined nitroglycerin and nitrosubstitution compound may also be determined, the nitrogen of the nitroglycerin deducted and the amount of nitrosubstitution compound calculated from the resulting difference, if the identity of the nitrosubstitution compound has been established. A suitable modification of the Kjeldahl method which has been found applicable to difficultly decomposable nitrocompounds is as follows:[1] This method is, of course, applicable to mixtures containing nitroglycerin.

Modified Kjeldahl Method for Nitrogen. About 0.5000 g. of the nitrocompound is weighed into a 500 cc. Kjeldahl flask, 30 cc. of 96% H_2SO_4 and 2 g. salicylic acid added and the sample dissolved by heating on a steam bath if necessary. Cool; add 2 g. zinc dust in small portions, with cooling and rotating the flask. Continue the shaking at 15 minute intervals for 2 hours and let stand overnight. Then heat over a small flame till fuming has ceased (about 2 hours), cool slightly and add 1 g. HgO. and boil 1–1½ hours longer. Cool and add 7.5 g. K_2SO_4 and 10 cc. H_2SO_4 and boil 1½ to 2 hours more. If the solution is not clear and almost colorless, add 1 g. more K_2SO_4 and boil longer. Cool and add 250 cc. H_2O to dissolve the cake formed, then add 25 cc. K_2S solution (80 g. per liter H_2O), 1 g. granulated Zn, and 85–90 cc. NaOH solution (750 g. per liter H_2O), and distill as usual in the Kjeldahl determination, collecting the NH_3 in standard H_2SO_4 solution. A blank determination without sample is advisable.

Separation of Nitrocompounds from Nitroglycerin. Hyde has devised a satisfactory method for actual separation of nitrosubstitution compounds from nitroglycerin, depending on the differences in solubility of these ingredients in carbon bisulphide and dilute acetic acid.[2] Nitroglycerin is only slightly soluble in CS_2, but readily soluble in dilute acetic acid, while most nitrocompounds are much more soluble in CS_2 and much less soluble in dilute acetic acid than nitroglycerin. CS_2 and acetic acid are only slightly miscible. Hence nitroglycerin and a nitrocompound may be partly separated by shaking the mixture with CS_2 and dilute acetic acid, allowing the two solvents to separate into two layers and drawing off one of the layers. The CS_2 layer will contain most of the nitrocompound and the acetic acid layer most of the nitroglycerin.

Hyde's method involves a continuous fractional extraction in a rather complicated apparatus consisting of 13 long narrow extraction tubes, connected with each other and with a condenser, reservoir and distilling flask so as to form a closed circulating system, the CS_2 continually passing in a train of fine drops through acetic acid in the series of extraction tubes, carrying with it the nitrocompound, the nitroglycerin tending to remain dissolved in the acetic acid. Practically a complete separation is finally obtained, the nitrocompound dissolved in the CS_2 collecting in the distilling flask at the end of the extraction train and the nitroglycerin remaining in solution in the acetic acid in the tubes. The CS_2 is evaporated and the nitrocompound weighed. Reference should be made to the original article by Hyde for details as to construction and operation of the apparatus.

[1] Cope, W. C., "Kjeldahl Modification for Determination of Nitrogen in Nitrosubstitution Compounds," J. Ind. and Eng. Chem., Vol. 8, p. 592, 1916.
[2] Hyde, A. L., "The Quantitative Separation of Nitrosubstitution Compounds from Nitroglycerin," J. Am. Chem., Soc., Vol. 35, p. 1173, 1913. (See also Bu. Mines Bulletin, 96, pp. 47–50, 1916.)

Nitrosugars. The nitrates of sugar, improperly called nitrosugar, are used to a considerable extent for lowering the freezing point of nitroglycerin. This substance is soluble in nitroglycerin, being prepared with the latter by nitrating a solution of cane-sugar in glycerin, and no method is known for its separation from nitroglycerin. Hoffman and Hawse[1] have reported on an optical method for the determination of nitrated sugar in nitroglycerin mixtures, based on the use of the polariscope. As an example of the application of the method, 10.65 g. of a nitrated mixture of glycerin and sugar was dissolved in 100 cc. alcohol and its angle of rotation found to be $a = 3.07°$. The specific rotatory power of sucrose octanitrate having been determined as $\alpha = 56.66$, the formula: C (concentration) $= a/2\alpha$ gives a result of 25.44% sucrose octanitrate in the sample.

The result of the optical method may be roughly checked by a determination of the total nitrogen of the combined nitroglycerin and nitrosugar, assuming the nitrogen content of the nitrosugar to be 15% (theoretical for sucrose octanitrate 15.95%), and that of nitroglycerin 18.50%.

Nitrochlorhydrins. Dinitromonochlorhydrin has been known for years as a partial substitute for nitroglycerin in explosives. It is a solvent for nitrocellulose in smokeless powders and has an appreciable effect in lowering the freezing point of nitroglycerin. During recent years it has come into use in this country as a substitute for nitrotoluenes in low freezing dynamites.

A mixture of dinitrochlorhydrin and nitroglycerin will have a lower nitrogen content than pure nitroglycerin, the dinitrochlorhydrin containing only 14.0% N, as compared with 18.50% N in nitroglycerin. The dinitrochlorhydrin may be readily identified and determined quantitatively by treating the mixture containing this substance and nitroglycerin with an excess of alcoholic solution of KOH, heating on the steam bath until saponification is complete, and determining the chlorine in the solution as chloride.

It must be noted that dinitrochlorhydrin is somewhat more volatile than nitroglycerin and therefore in evaporating the ether from the ether extract it is advisable to make use of the bell-jar evaporator (p. 1376) so as to minimize its loss during evaporation.

Nitropolyglycerin. Nitrated polymerized glycerin—usually a mixture of tetranitrodiglycerin and trinitroglycerin—is sometimes found in low-freezing explosives. This mixture will show a lower N-content than nitroglycerin, since pure tetranitrodiglycerin contains only 16.19 % N. The presence of the latter substance is indicated by low solubility in dilute acetic acid (60 volumes glacial acetic acid to 40 volumes water). One gram of nitroglycerin dissolves in about 10.5 cc. of this acid, while 1 gram of a mixture containing 82.25% tetranitrodiglycerin required 120 cc. of the acetic acid to completely dissolve it. In dissolving such a mixture, it will be found that a part of the mixture dissolves more readily than the remainder. If the more difficultly soluble portion is separated, dried in a desiccator and its nitrogen content determined in the nitrometer, it will be found to contain a much lower % N than the original mixture, approximating the figure for tetranitrodiglycerin, 16.19% (an actual trial gave 16.24% N).

If the presence of tetranitrodiglycerin is established by the above procedure and no other substances except nitroglycerin are present, the proportions of these two ingredients in the ether extract may be readily calculated from the N-content as found by the nitrometer.

[1] Hoffman, E. J. and Hawse, V. P., "The Nitration of Sucrose Octanitrate," J. Am. Chem. Soc., Vol. 41, pp. 235–247, 1919.

" PERMISSIBLE " EXPLOSIVES

"Permissible" explosives are coal mining explosives which have passed the prescribed tests of the Bureau of Mines and are recommended by the Bureau for use in gassy and dusty mines. Their important characteristic is a relatively low flame temperature, which is brought about by modifying the composition of the usual types of dynamites and other blasting explosives. The general methods of reducing the flame temperature of explosives[1] are summarized as follows:

(a) Addition of an excess of carbon,—forming less CO_2 and more CO in the gases of explosion.
(b) Addition of free water or of solids with water of crystallization.
(c) Addition of inert materials.
(d) Addition of volatile salts.

The analysis of explosives of this class is therefore generally more complicated than that of the ordinary types of blasting explosives because of the greater variety of ingredients used in manufacture. A partial list of substances which have been found in low-flame explosives manufactured in this country is shown below, arranged according to their solubility in the general scheme of analysis:

Soluble in Ether
Nitroglycerin
Nitropolyglycerin
Nitrotoluenes
Nitrosugars
Nitrochlorhydrins
Paraffin
Resins
Sulphur
Vaseline
Oils

Soluble in Water
Ammonium nitrate
"   chloride
"   sulphate
"   oxalate
"   perchlorate
Alum (cryst.)
Aluminum sulphate (cryst.)
Barium nitrate
Calcium sulphate (cryst.)
Gums
Magnesium sulphate (cryst.)
Potassium chlorate
"   nitrate
"   perchlorate
Sodium nitrate
"   chloride
"   bicarbonate
"   carbonate
Sugar
Zinc oxide

Soluble in Acids
Aluminum
Calcium carbonate
"   silicide
Ferric oxide
Magnesium carbonate
Zinc
Zinc oxide

Insoluble

Charcoal
Clay
Coal
Corn meal
Corncob meal
Kieselguhr
Nitrocellulose
Nitrostarch
Nitrated wood

Peanut shell meal
Powdered slate
Rice hulls
Sawdust
Turmeric
Vegetable ivory meal
Wheat flour
Wood pulp

[1] The thermochemical considerations involved are discussed in Bureau of Mines Bulletin No. 15, "Investigations of Explosives used in Coal Mines," 1912, and the details of analysis in Bureau of Mines Bulletin No. 96, "The Analysis of Permissible Explosives," C. G. Storm, 1916.

Qualitative Analysis. The qualitative examination of a "permissible" explosive is conducted in the same manner as has been described for dynamite (see page 1374), and, in view of the greater variety of constituents that may be present, is quite essential before a suitable scheme for quantitative separation can be chosen.

Tests for some of the more unusual substances not generally found in the ordinary types of blasting explosives, and not already discussed under "Low-freezing Explosives," are made as follows:

Test for Sugar. The presence of water-soluble organic substances is indicated by an appreciable charring of the residue obtained by evaporating a portion of the water extract to dryness and then heating gradually over a burner. A slight charring may result from water-soluble portions of cereal products, wood-pulp, etc., and may be disregarded. Sugar is identified by acidifying some of the water solution with a little dilute HCl, heating to boiling, neutralizing with KOH and then boiling with Fehling's solution. A precipitation of cuprous oxide indicates the presence of sugar.

Test for Gum Arabic. Gum arabic is precipitated by the addition of a solution of basic lead acetate to the water extract, a white, flocculent precipitate of indefinite composition resulting (see Determination of Gum Arabic, p. 1388).

Test for Nitrostarch. Nitrostarch is best identified by microscopic examination of the residue insoluble in water. It is easily distinguished from unnitrated starch by means of a solution of iodine in KI, which colors the starch granules dark blue but does not affect the granules of nitrostarch.

Test for Chlorides, Chlorates, and Perchlorates. These three substances present in a solution may be identified as follows: Acidify slightly with nitric acid, add excess of $AgNO_3$, heat to boiling, shake well, and filter off the silver chloride. To the filtrate add a few cc. of 40% solution of formaldehyde (formalin), and boil to reduce chlorates to chlorides. This reduction is best carried out by heating on the steam bath for about an hour. Any chloride thus formed is then separated by further precipitation with $AgNO_3$ and removed by filtration. The filtrate is then evaporated to dryness, the residue transferred to a crucible and fused with dry Na_2CO_3. The fused mass is treated with dilute HNO_3, when the presence of perchlorate will be indicated by an insoluble precipitate of AgCl.

Mechanical Separation of Solid Ingredients. It is frequently of advantage, especially in connection with the interpretation of the results of analysis of an explosive mixture containing a number of water-soluble salts, to determine the identity of one or more of the components of the mixture by means of screening or by a method of separation depending on variation in specific gravity of the components. Such methods are facilitated by the fact that the ingredients of blasting explosives are frequently not finely pulverized in the course of manufacture.

(a) *By Screening.* 25 to 50 grams of the sample is washed several times with ether to remove nitroglycerin and ingredients of an oily nature, the solid residue dried to remove adhering ether and then sifted through a set of sieves. An examination of the portions held by the 10- and 20-mesh screens will usually show the presence of coarse crystals which are large enough to be sorted out with the aid of forceps, submitted to qualitative tests and identified with certainty. A single crystal may sometimes be identified by dissolving it in a drop of water on a microscope slide, allowing the water to evaporate and examining the resulting crystals under the microscope. The writer has frequently identified three or four ingredients of an explosive in this manner.

(b) *By Specific Gravity Separations.* This method, applied to the analysis of explosives by Storm and Hyde,[1] depends on the separation of solids from a mixture by means of inert liquids of different specific gravities. A series of mixtures of chloroform (sp.gr. 1.49) and bromoform (sp.gr. 2.83) is prepared covering as wide a range of specific gravity as may seem desirable. Portions of the dried sample previously extracted with ether as in (a) are added to such liquid mixtures and the heavier salts, which settle to the bottom, separated from the lighter ones. For example a mixture of ammonium nitrate (sp.gr. 1.74) and sodium chloride (sp.gr. 2.17) is readily separated into its components in a liquid with a specific gravity of (e.g.), 1.90, so that the components can be tested separately and the analyst assured that the mixture is not composed of sodium nitrate and ammonium chloride,—which could not be ascertained by ordinary quantitative analysis. (For example, a mixture composed of 16.61% Na, 44.76% NO_3, 13.00% NH_4 and 25.63% Cl may contain either 61.37% $NaNO_3$ and 38.63% NH_4Cl, or 57.76% NH_4NO_3 and 42.24% NaCl, or varying proportions of all four ingredients.) The chloroform-bromoform mixtures are recovered by filtering and used repeatedly.

The specific gravities of some of the more common salts that may be found are as follows:

```
Ammonium alum (cryst.)...........................1.62
   "      chloride..................................1.52
   "      nitrate...................................1.74
   "      perchlorate...............................1.87
   "      sulphate..................................1.77
Barium nitrate....................................3.23
Calcium carbonate (ppt'd).........................2.72
   "      sulphate (anhydrous)......................2.97
   "      sulphate+2H2O.............................2.32
Magnesium carbonate...............................3.04
   "      sulphate+7H2O.............................1.68
Manganese dioxide.................................5.03
Potassium alum (cryst.)...........................1.75
   "      chlorate..................................2.33
   "      chloride..................................1.99
   "      nitrate...................................2.09
   "      perchlorate...............................2.52
   "      sulphate..................................2.66
Sodium chloride...................................2.17
   "      nitrate...................................2.26
   "      sulphate (anhydrous)......................2.66
   "      sulphate+10H2O............................1.46
```

Moisture. The determination of moisture in all types of "permissible" explosives is carried out by the method described for nitroglycerin dynamites (page 1375). The influence of the slight volatility of nitroglycerin and of certain nitrosubstitution compounds on the results of this determination has been discussed (pp. 1375, 1381). A more serious factor in the case of many "permissible" explosives is the presence of salts containing water of crystallization. Most salts of this type (e.g., $MgSO_4.7H_2O$) undergo a gradual loss of a large part of their combined water on desiccation over either H_2SO_4 or $CaCl_2$, thus rendering it impossible to differentiate between hygroscopic moisture and

[1] Storm, C. G., and Hyde, A. L., "Specific Gravity Separation Applied to the Analysis of Mining Explosives," Tech. Paper No. 78, Bureau of Mines, 1914.

combined water. Attempts to remove the total water content by heating at a temperature high enough to drive off all of the water of crystallization are useless on account of the increased volatilization of nitroglycerin, ammonium nitrate, etc., at such temperatures.

In such cases it is necessary to determine all other constituents by direct methods and estimate moisture by difference, the salt to which the water of crystallization belongs being calculated as containing its full quota of water; or the crystallized salt may be calculated as anhydrous and the difference from 100% reported as "water of crystallization plus moisture."

Extraction with Ether. The extraction with ether, the evaporation of the ether, and the analysis of the ether-soluble portion are conducted as already discussed for nitroglycerin dynamites (pp. 1375, 1376).

In drying the crucibles containing the residue insoluble in ether, a temperature of 100° C. may be used except when the residue contains ammonium nitrate or organic nitrates such as nitrocellulose, nitrostarch, or nitrated wood. When any of these substances are present, the residue should be dried to constant weight at 70° C. Except when salts containing water of crystallization are present, the amount of ether-soluble material found is calculated by deducting the moisture determined by desiccation, from the difference between the weight of original sample and the weight of the dried residue insoluble in ether. The procedure followed when water of crystallization is present is noted in the preceding paragraph.

Extraction with Water. Water-soluble salts are extracted from the weighed residue insoluble in ether as already described, the residue left in the crucibles dried to constant weight at 95 to 100° C., cooled and weighed. The water soluble salts in the solution are determined by the usual methods of inorganic analysis.

Nitrates. In determining nitrates by the nitrometer method (see p. 354) it must be remembered that the presence of a considerable quantity of chlorides may interfere with the accuracy of the results. Many of the "permissible" explosives contain sodium chloride in amounts varying from 1% to 10 or 15%. M. T. Sanders[1] has shown that if the sodium chloride is present in an amount exceeding 15–17% of the sodium nitrate, the result is not accurate within 0.1%. Smaller amounts of sodium chloride do not interfere, except to increase the amount of sludge formed in the nitrometer.

Nitrates may also be determined by the "nitron" method of Busch.[2]

Chlorates. Chlorates may be determined by any of the methods described on page 152 (reduction with SO_2, $FeSO_4$, or Zn) or by the formaldehyde method.[3] In the latter method a portion of the solution, containing about 0.5 g. of chlorate is diluted to 150 cc., 5–10 c. of 40% formaldehyde solution, 2 cc. dilute HNO_3 (1 : 3), and 50 cc. of approx. tenth normal silver nitrate added, the solution covered and heated on the steam bath for about 4 hours, when the precipitate of AgCl is filtered off, washed, dried and weighed. This method is accurate to .05 to .10%.

[1] Sanders, M. T., "The Effect of Chlorides on the Nitrometer Determination of Nitrates," J. Ind. & Eng. Chem., 12, p. 169–170, 1919.
[2] See page 345, also, for further details, "The Analysis of Permissible Explosives," Bureau of Mines, Bulletin 96, pages 60–2.
[3] Storm, C. G., "The Analysis of Permissible Explosives," Bulletin 96, Bureau of Mines, pp. 63–4, 1916.

Perchlorates. The determination of perchlorates by reduction to chlorides on ignition with NH_4Cl in the presence of platinum is described on page 1128. Perchlorates may also be determined by means of precipitation with "nitron" in exactly the same manner as for nitrates. The weight of nitron perchlorate $(C_{20}H_{16}N_4HClO_4)$ found, multiplied by 117.5 (mol. wt. of NH_4ClO_4) and divided by 412.5 (mol. wt. of nitronperchlorate) gives the weight of perchlorate found, expressed as NH_4ClO_4.

Gum Arabic. This substance, sometimes used as a binder in dry explosive mixtures—especially chlorate or perchlorate powders—is determined by precipitation with basic lead acetate solution, prepared by adding 150 g. of normal lead acetate and 50 g. lead oxide (PbO) to 500 cc. distilled water, heating almost to boiling, and filtering. This reagent is added to the solution containing the gum arabic until no further precipitation occurs; the mixture is allowed to stand for several hours, then filtered, washed with absolute alcohol, dried at 100° and weighed. The weight of precipitate multiplied by the factor 0.4971 (determined experimentally) gives the weight of gum arabic found. Chlorides or sulphates, if present, interfere with the determination and must be first removed.

Sugar. Sugar may be present as an ingredient in some "permissible" explosives, and is always found in small amounts in the water extract if cereal products such as corn meal or wheat middlings are present. A portion of the water extract is acidified with HCl (1 cc. conc. HCl to 100 cc. solution), heated just to boiling, cooled, nearly neutralized with Na_2CO_3, an excess of Fehling's solution added and the mixture heated until reduction is complete. The Cu_2O is filtered from the blue liquid, dried, ignited to constant weight, and weighed as CuO. This weight$\times$0.4308 equals weight of cane sugar. The result is corrected for the result of a blank determination using distilled water instead of the water extract. By the use of this method after first extracting with ether, then with water, corn meal was found to contain 2.65% and wheat middlings 6.25–7.00% of sugar. Thus an explosive containing 25% wheat middlings would show as much as 1.75% of sugar in its water extract.

Extraction with Acid. As in the ordinary nitroglycerin dynamites, the substances removed from "permissible" explosives by acid extraction are chiefly substances added as antacids, including calcium carbonate, magnesium carbonate and zinc oxide. Other acid-soluble materials that may be present include metallic aluminum or zinc, ferric oxide, manganese dioxide, and calcium silicide. When starch is present, the residue from the water extraction is subjected to hydrolysis in boiling dilute HCl as already described (page 1378), and the acid-soluble inorganic components determined in the filtrate by the usual methods. An extraction with cold acid is made only when there is no starch present.

Extraction with Acetone: Nitrocellulose and Nitrostarch. If either nitrocellulose or nitrostarch is present, an extraction with acetone is made as described for gelatin dynamite (page 1401). It should be noted in connection with the preceding steps in the analysis that in order to avoid partial solution of these substances in ether, the ether used in the ether extraction should be alcohol-free (distilled from sodium), and also that all drying of residues containing these materials should preferably be conducted at 70° instead of 100°, in order to avoid partial decomposition. It is impracticable to separate nitrostarch from nitrocellulose but they are not likely to be found together in the same explosive. Small amounts of nitrocellulose are detected less readily than nitrostarch, which is easily identified by the microscope.

Insoluble Residue and Ash. The insoluble residue is usually carbonaceous combustible or absorbent material and is in most cases readily identified by means of a microscope (preferably binocular) with low power (25–50 diameters). The possible presence of any inorganic material which may have been over-looked in the analysis is detected by means of a determination of ash, the residue being ignited until all carbon is burned off and the mineral residue weighed.   This is usually not over 0.2%.   If higher than 0.5%, there is reason to suspect that some such material as kieselguhr or clay is present, or that the extractions with water or acid were not complete.

NITROSTARCH EXPLOSIVES

General Nature. Nitrostarch explosives have been for a number of years used to a very considerable extent in this country for commercial blasting purposes, chiefly for quarrying.   During the war, explosives of this class were adopted by the United States for certain military purposes and proved satis-factory substitutes for trinitrotoluene as bursting charges for hand grenades, rifle grenades and trench mortar shell.

The commercial nitrostarch explosives may contain, in addition to the nitrostarch, any or all of the following components: oxidizing agents, as sodium or ammonium nitrates, combustible material, such as charcoal, flour, sulphur, etc., mineral oil, and antacids, such as calcium carbonate, or zinc oxide.   Nitrostarch military explosives may consist of some such mixture as the above, or may be composed almost entirely of nitrostarch with the addition of relatively small amounts of oils and materials used for granulating.   Some types of nitrostarch explosives proposed for military use contained water in amounts up to 10–15% combined with a mixture of nitrostarch and a soluble nitrate.

For commercial use, nitrostarch explosives are put up in cartridge form in paper wrappers, just as nitroglycerin explosives are prepared.

Moisture. Moisture is determined as in the case of nitroglycerin dyna-mites.   3 to 5 grams of the sample is desiccated over sulphuric acid for 2–3 days (2 days is usually sufficient to give constant weight), or for 24 hours in a vacuum desiccator with a vacuum of at least 700 mm. of mercury.   Unlike nitroglycerin, nitrostarch does not volatilize and may be desiccated to con-stant weight.   When heated to higher temperatures, e.g., 100° C., for any extended time, it undergoes, like all nitric esters, a gradual decomposition with loss of weight.

Extraction with Petroleum Ether: Oils, Sulphur, etc. The small amount of alcohol usually present in ordinary grades of ethyl ether, is sufficient to cause partial solution of the nitrostarch, if ethyl ether is used for removing oily ingredients, sulphur, etc., nitrostarch being readily soluble in mixtures of ether and alcohol.   It is therefore advisable to use petroleum ether, which does not dissolve nitrostarch, for this extraction.

A sample of about 10 grams of the explosive, in a Gooch crucible with asbestos mat, is extracted with pure petroleum ether of about 0.65 specific gravity, the excess solvent removed by suction and the crucible with sample dried to constant weight at approximately 70° C. The % loss of weight, minus the moisture content, already determined, represents the percentage of ether-soluble material present. The petroleum ether is removed by evaporation, the residue of ether-soluble materials dried, weighed, and its components determined by the methods used for dynamites.

Extraction with Water: Nitrates, Gums, etc. The dried and weighed residue insoluble in petroleum ether is extracted with distilled water to remove the nitrates or other water-soluble materials. The insoluble residue left in the crucible is dried at a temperature of 80° C. (100° may cause some decomposition of the nitrostarch) for several hours, and weighed, the loss of weight being total water extract and serving as a check against the sum of the components separately determined.

Ammonium nitrate, if present, is determined in the water solution by the usual method of distillation of the NH_3 after adding an excess of alkali. Sodium nitrate is determined by evaporating the water extract, volatilizing ammonium salts, and weighing the residue after re-oxidizing with nitric acid if charring has indicated the presence of any water-soluble organic material (page 1378).

If the original explosive is granular in form, the presence of a binding or agglutinating material in the water extract may be suspected. Although numerous other substances may be used for the purpose, gum arabic is frequently employed as a binding agent in different types of dry explosives. A qualitative test for gum arabic has been mentioned on page 1385, and its quantitative determination may be conducted as described on page 1388, by precipitation with basic lead acetate solution.

Insoluble Residue: Nitrostarch, Charcoal, Cereal Products, etc. *Starch.* A microscopic examination of the weighed insoluble residue will usually serve to identify its components. Any un-nitrated starch or cereal products is readily distinguished from nitrostarch by treating with a drop of KI solution of iodine and examining under the microscope, when the un-nitrated starch granules will appear blue or black and the nitrated starch colorless or yellow. Charcoal is identified by its color.

Un-nitrated starch, if present in an amount greater than a trace, is determined by boiling with dilute H_2SO_4 or HCl until iodine solution no longer colors a drop of the liquid blue, then filtering and washing thoroughly. The residue is dried at 100° and weighed, the loss of weight representing starch.

Nitrostarch. Another portion of the insoluble residue (the analysis being conducted in duplicate), is extracted with acetone in a Wiley extractor or other continuous extraction apparatus, or by transferring the residue from the crucible to a small beaker, digesting in acetone with stirring, and filtering through the same crucible, washing with fresh acetone. This extraction dissolves all of the nitrostarch, leaving any charcoal or cereal products that may be present. The residue is dried at 100° and weighed, the loss of weight representing nitrostarch.

The nitrogen content of the nitrostarch may be determined, if desired, by precipitating the nitrostarch from the clear acetone solution by the addition of water and evaporation on a steam bath. A portion of the white, floury precipitate is then dried at 70°–80° C., weighed, and its nitrogen content determined in the nitrometer (see page 354 Vol. I).

Charcoal. If charcoal is present its weight may be taken direct, in the absence of cereal products or other substances. When the residue contains cereal products, the material left after hydrolysis of the starch and extraction of the nitrostarch will contain the crude fibre of the cereal together with charcoal or other insoluble ingredients. A separation of such components is usually impracticable.

Trinitrotoluene (TNT.)

Trinitrotoluene, commonly designated in this country by the abbreviation TNT., is also known in this and other countries by such names as triton, trotyl, tolite, trilite, trinol, tritolo, etc. The term trinitrotoluol, which is probably more commonly used than trinitrotoluene, is incorrect according to approved chemical nomenclature.

This explosive is of the greatest importance as a high explosive for military use, being adaptable as a bursting charge for high explosive shell, trench mortar shell, drop-bombs, grenades, etc., because of its powerful explosive properties, relative safety in manufacture, handling, etc., its stability, its lack of hygroscopicity, and absence of any tendency to form sensitive compounds with metals.

It is classified by the Ordnance Department, U. S. A., into three grades, according to purity—Grades I, II, and III, with solidification points of at least 80.0°, 79.5°, and 76° C., respectively. Other requirements—the same for all grades—are as follows: ash, not more than 0.1%; moisture, not more than 0.1%; insoluble, not more than 0.15%; acidity, not more than 0.01%.

Solidification Point. The determination of the solidification point or "setting point" of TNT. is the best single test for purity of this compound, and is preferably carried out as follows:

A sample of about 50 grams of TNT. is placed in a 1″×6″ test tube and melted by placing the tube in an oven at about 90° C. The tube is then inserted through a large cork stopper into a larger test tube about $1\frac{1}{2}″×7″$, which, in turn, is lowered into a wide-mouth liter bottle, so that the rim of the large tube rests on the neck of the bottle. The inner test tube is provided with a cork stopper containing 3 openings—one for a standard thermometer graduated in 1/10° C., one for a short thermometer which is passed just through the stopper and is used for noting the average temperature of the exposed mercury column of the standard thermometer, and the third opening being a small v-shaped notch at the side of the stopper, through which passes a wire whose lower end is bent in a loop at right angles to the axis of the tube and which is used for stirring the molten sample of TNT.

The standard thermometer is so adjusted that its bulb is in the center of the molten mass, and the stirrer is operated vigorously, the thermometer being watched carefully as the temperature falls. The temperature will finally remain constant for an appreciable time and then rise slightly, owing to the heat of crystallization of the TNT. As this point is reached, readings should be taken about every 15 seconds until the maximum temperature of the rise is reached. This temperature will usually remain constant for several minutes while crystallization is proceeding. The maximum reading, corrected for the emergent stem of the thermometer, is taken as the solidification point of the sample.

Ash. About 5 grams of TNT. is moistened with sulphuric acid and burned in a tared crucible. The residue is again moistened with a few drops of nitric acid and sulphuric acid and again ignited and the resulting ash weighed.

Moisture. A sample of about 5 grams spread on a watch glass is desiccated over sulphuric acid to constant weight.

Insoluble. A sample of about 10 grams is treated with 150 cc. of 95% alcohol, heated to boiling, and filtered while hot through a tared Gooch crucible with asbestos mat. The insoluble residue is washed with hot alcohol, dried at 100° C. and weighed.

Acidity. A 10-gram sample is melted in a large test tube or a flask and shaken with 100 cc. of neutralized boiling water, cooled and the water decanted. A similar treatment is given using 50 cc. of boiling water, the two portions of water combined, cooled and titrated with tenth normal NaOH, using phenolphthalein indicator. The acidity is calculated as % H_2SO_4 in the original sample.

Nitrogen. Nitrogen is not usually determined in the inspection of TNT. but when necessary it may be determined by the Dumas combustion method or the modification of the Kjeldahl method described on page 1382.

Picric Acid

Ordnance Department, U. S. A., specifications for picric acid prescribe that it shall have a solidification point of not less than 120° C.; that it shall contain not more than the following amounts of impurities:

Moisture—0.2% for dry material.................12.0% for wet.
Sulphuric acid (free and combined)............... 0.10%
Ash... 0.2%
Insoluble in water............................. 0.2%
Soluble lead................................... 0.0004%
Nitric acid (free)............................. none

Solidification Point. Dry the sample at a temperature not exceeding 50° C. Melt sufficient to give a 3-inch column in a 6-inch×¾-inch test tube immersed in a bath of glycerin heated to 130° C. When the sample is completely melted remove the tube from the bath and stir the sample with a standardized thermometer graduated in 0.10 degrees, until the picric acid solidifies. During solidification the temperature will remain constant for a short time and then undergo a slight rise. The highest temperature reached on this rise is recorded as the solidification point. The test may be more accurately carried out using the apparatus and method as described under trinitrotoluene (p. 1391).

Moisture. A weighed sample of about 10 grams is spread evenly on a tared watch glass and dried to constant weight (about 3–4 hours) at 70° C.

Sulphuric Acid. About 2 grams is weighed and dissolved in 50 cc. of distilled H_2O, acidified with HCl and heated to about boiling. Hot $BaCl_2$ solution is added with stirring and the mixture allowed to stand at least 1 hour on the steam bath. Filter hot on a tared Gooch crucible, wash with water, dry at 100° C. and weigh. Calculate $BaSO_4$ found as H_2SO_4 in original sample.

Ash. About 1 gram is weighed in a platinum crucible, saturated with melted paraffin, burned carefully, and the residue ignited to burn off all carbon. The resulting ash is cooled and weighed.

Insoluble in Water. 10 grams of the sample is treated with 150 cc. boiling water, boiled for 10 minutes, filtered while hot through a tared Gooch crucible, washed well with hot water, and the insoluble residue on the filter dried at 100°, cooled, and weighed.

Soluble Lead. The presence of soluble lead in picric acid is highly objectionable, because lead picrate is an extremely sensitive explosive and its presence would greatly increase the dangers involved in handling and loading picric acid. A weighed sample of about 300 g. is digested in a 2-liter flask with 100 cc. of a hot saturated solution of barium hydroxide in 65% alcohol. 1400 cc. of 95% alcohol is then added and the digestion continued at a temperature below the boiling point (with reflux condenser), until everything except traces of insoluble matter is in solution. The picric acid is then allowed to crystallize on cooling, and the solution filtered off, decanting the clear liquid from the crystals until 500 cc. of filtrate is obtained. This 500 cc., representing 100 g. of picric acid, is treated with 5 drops HNO_3 and 10 cc. of 1% $HgCl_2$ solution, and H_2S passed through it for 15 minutes. Allow the precipitate to settle for 20 minutes, filter and wash with alcohol saturated with H_2S. Dry and ignite the precipitate, then dissolve the residue in 9 cc. of HNO_3 (sp.gr. 1.42) by warming, add warm water to bring the volume to 50 cc., and electrolize at 0.4 ampere and 2.5 volts, temperature 65° C., for 1 hour. Wash the electrode by replacing the beaker with another one containing distilled water without interrupting the current. Dry and weigh the previously tared anode. The weight of lead peroxide found $\times 0.8661$ gives the percentage of soluble lead found.

Nitric Acid. No coloration should result when a water solution of picric acid is treated with a solution of diphenylamine in sulphuric acid.

Ammonium Picrate

Ammonium picrate, also known in this country as "Explosive D," is of importance as a military explosive more on account of its insensitiveness to shock and friction, than because of its explosive strength, which is less than that of TNT. Its chief use is as a bursting charge in armor-piercing projectiles.

Military specifications require it to be prepared from picric acid of standard purity, to contain not less than 5.60% ammoniacal nitrogen, and not more than the following amounts of impurities:

Moisture	0.20%
Sulphuric acid (free and combined)	0.10%
Nitrates	trace
Insoluble material	0.20%
Ash	0.20%
Nitrophenols	0.50%

Moisture. A sample of about 10 grams spread on a tared watch glass is dried at 95° C. to constant weight (about 2 hours).

Sulphuric Acid. About 5 grams is dissolved in 100 cc. of hot water, filtered, washed with 25 cc. hot water, the filtrate acidified with HCl, heated to boiling and treated with hot $BaCl_2$ solution. Any precipitate is filtered on a weighed Gooch crucible, dried at 100° C., weighed and calculated as H_2SO_4.

Nitrates. A water solution of the sample tested with diphenylamine and H_2SO_4 should give no blue coloration.

Insoluble Material. A 10-gram sample is boiled with 150 cc. of water for 10 minutes, filtered on a Gooch crucible, the residue washed with hot water, dried at 100° and weighed.

Ash. A sample of about 1 gram is saturated with melted paraffin and burned in a tared crucible, the residue ignited to burn off all carbon, and the ash weighed.

Nitrophenols. 10 grams of powdered sample is treated with 50 cc. of chloroform for 30 minutes with frequent stirring and filtered into a 100 cc. tared flask, the residue being washed with 25 cc. of chloroform. The filtrate is evaporated to dryness and any residue obtained weighed. This residue is treated with ammonium hydroxide, again evaporated to dryness and extracted the second time with 25 cc. of chloroform. The chloroform filtrate is evaporated to dryness and the residue weighed. The difference in weight between this residue and the first residue equals the nitrophenols, other chloroform-soluble having been eliminated by the ammonia treatment and second extraction.

Tetryl

Tetryl is the commercial term applied to the explosive trinitrophenyl-methylnitramine, also improperly called tetranitromethylaniline. Its chief use is as a "booster" charge in high explosive shell, where it serves to transmit the detonating wave from the detonator or fuze to the less sensitive bursting charge. Being in immediate contact with the fuze it must be of a high degree of purity, and is required by Ordnance Department specifications to have a melting point of at least 128° C. and to contain not more than the following amounts of impurities:

Moisture...0.05%
Acidity (as H_2SO_4)....................................0.01%
Insoluble in acetone....................................0.30%
Ash...0.15%
Sodium salts..trace

Melting Point. The sample to be used for this test is dried overnight in a vacuum desiccator and pulverized to pass a 100-mesh screen. A capillary melting-point tube is filled to about $\frac{1}{4}$ inch from the bottom and attached to the stem of a standard thermometer so that the sample is next to the center of the bulb. The bath is properly agitated and provision made for correcting for the emergent stem of the thermometer. The temperature of the bath is raised rapidly to 120° C., then at the rate of 1° in 3 minutes, the temperature at which the first meniscus appears across the capillary tube being noted as the melting point.

Moisture. A sample of about 10 grams is weighed in a wide shallow weighing bottle and dried over sulphuric acid in a desiccator for 24 hours, the sample being spread uniformly so that its depth is not over 0.5 cm. The loss of weight is regarded as moisture.

Acidity. A 10-gram sample, finely powdered, is shaken for 5 minutes with 50 cc. of boiled distilled water, filtered, washed with 50 cc. more water, and the filtrate and washings titrated with N/50 NaOH solution using phenolphthalein indicator.

Insoluble in Acetone. 10 grams of sample is dissolved in 75 cc. of acetone, filtered through a tared Gooch crucible, and the residue washed with 25 cc. of acetone, dried to constant weight at 100° C. and weighed.

Ash. The dried residue insoluble in acetone is ignited, cooled in a desiccator and weighed.

Sodium Salts. Any sodium present in tetryl is combined as sodium picrate. 10 grams of the tetryl are boiled in 50 cc. distilled water, cooled, filtered, the filtrate acidified with acetic acid and evaporated to 10 cc., cooled again and filtered. The filtrate is made alkaline with ammonia and treated with 5 cc. of 10% solution of ammoniacal copper sulphate. Any sodium picrate will be precipitated as crystals of cupro ammonium picrate on standing for a few minutes.

Mercury Fulminate

In commercial blasting caps and electric detonators mercury fulminate is generally found intimately mixed with potassium chlorate. It is, however, used without admixture in certain types of detonators, in the fuzes of high explosive shell and for other military purposes. It is usually purchased under specifications which provide that it shall be at least 98% pure, shall be free from acid, and contain not more than 2% insoluble matter, 1% free mercury, and 0.05% chlorine in the form of chlorides.

Preparation of Sample. Mercury fulminate being packed and handled in a thoroughly wet condition until dried just before use, it is generally necessary to dry the sample before testing. This may be done by exposing in a low temperature oven at not more than 50° C. until practically dry, then in a desiccator (not a vacuum desiccator) over sulphuric acid or calcium chloride until its weight is constant.

Mercury Fulminate Content. Exactly 0.3 g. is weighed into a wide-mouthed Erlenmeyer flask containing 250 cc. distilled water, and 30 cc. of a 20% solution of purest sodium thiosulphate is added quickly and the mixture shaken for exactly 1 minute. At once titrate with N/10 hydrochloric acid using 3 drops of methyl orange indicator, the titration to be commenced 1 minute after adding the sodium thiosulphate, and to occupy not more than 1 minute additional time.

The percentage of mercury fulminate is calculated from the volume of standard acid required, after deducting the volume of acid required for a blank determination. Four molecules of HCl are equivalent to 1 mol. of mercury fulminate, or 1 cc. N/10 HCl equals 0.00711575 g. mercury fulminate. The reaction is assumed to be as follows:

$$HgC_2N_2O_2 + 2Na_2S_2O_3 + 2H_2O = HgS_4O_6 + 4NaOH + C_2N_2.$$

Acidity. A 10-g. sample is extracted with 2 successive 25-cc. portions of boiled distilled water in a Gooch crucible, and 3 drops of methyl orange solution (1 g. per liter) added. No red tinge of color should be obtained.

Insoluble Matter. A 2-g. sample is dissolved in hot 20% $Na_2S_2O_3$ solution, filtered through a tared Gooch crucible and any insoluble washed with water then with alcohol and finally with ether, dried at 60°–70° C. and weighed.

Free Mercury. The residue of insoluble matter obtained as described above is treated with a solution of 3 g. KI and 6 g. $Na_2S_2O_3$ in 50 cc. H_2O by passing the solution through the Gooch crucible. Any organic mercury compounds are thus converted into mercuric iodide, which is soluble in $Na_2S_2O_3$ solution. The metallic mercury remains behind on the filter, and is washed with H_2O, dried 1 hour at 80°–90° C., and weighed.

Chlorides. A 5-g. sample of fulminate is extracted in a Gooch crucible with 2 successive 25 cc. portions of distilled water at 90°–100° C. Three drops of strong HNO_3 and 10 drops of 10% $AgNO_3$ solution are added to the filtrate. If a turbidity results, the AgCl should be determined gravimetrically or a fresh sample extracted and the filtrate titrated with a standard $AgNO_3$ solution.

Blasting Caps and Electric Detonators

Preparation of Sample. In the examination of blasting caps or detonators for either commercial or military use, the removal of the detonating composition from the copper or brass shell requires considerable precaution. Blasting caps are emptied by squeezing the cap gently in a pair of "gas forceps," the jaws of the forceps being passed through a small opening in a piece of heavy leather, rubber belting, or similar material, about 6″ square, which serves as a shield to protect the hand in case of the possible explosion of the cap in squeezing. After each squeeze, the loosened portion of the charge is shaken out on a piece of glazed paper, the cap turned slightly in the forceps and again squeezed. The pressure on the cap should be just sufficient to slightly dent it, and in shaking out the charge, the cap should not be tapped on the table or other surface. With these precautions there is little danger of an explosion.

Electric detonators are opened by first cutting off the wires or "legs" close to the shell, then tearing off the upper portion of the shell by means of pointed side-cutting pliers, the cap being held firmly in the fingers and a thin strip of the copper shell being torn off spirally by nipping the top edge of the shell with the forceps. This must be done with great care, especially as the portion of the shell containing the fulminate charge is approached. When the greater portion of the plug which holds the wires in place has been exposed, the plug and wires are gently pulled out, care being taken to avoid force and possible friction, and any adhering particles of the charge brushed off onto glazed paper. The charge is then removed from the lower part of the shell just as in the case of blasting caps.

The charge is removed separately from several of the caps or detonators and each weighed in order to determine the average weight of charge as well as variation of same.

"Reinforced" caps, or those which contain a small perforated inner copper capsule pressed on top of the charge, must be opened in the manner described for electric detonators, in order to remove the inner capsule. Detonators of this type usually contain a main charge of some nitro compound superimposed by a layer of mercury fulminate, mixture of fulminate and chlorate, or lead azide. Although a clean mechanical separation of the two layers is usually not possible, portions can be taken from each and identified by qualitative tests before proceeding with a quantitative examination.

Moisture. The moisture content of the composition is determined by desiccating to constant weight over sulphuric acid or calcium chloride.

Analysis of Composition Containing Mercury Fulminate and Potassium Chlorate. About 2–3 grams of the well-mixed composition is weighed in a Gooch crucible provided with asbestos mat or disc of filter paper or silk, and first moistened with a few drops of alcohol, then extracted with 200–250 cc. of cold water in 15–20 cc. portions, using slight suction after each portion has remained in the crucible for a few minutes. The residue in the filter is dried to constant weight at 60°–70° C. (2–3 hours), and weighed.

The water extract contains the potassium chlorate and a portion of the mercury fulminate, which is slightly soluble in cold water. It is treated with 2 cc. of ammonium hydroxide and H_2S passed to completely precipitate the dissolved mercury fulminate as HgS. This black precipitate is filtered off, washed, dried and weighed. Its weight $\times 1.22$ gives the amount of mercury fulminate dissolved by the water. This weight added to the weight of the dried residue insoluble in water gives the total weight of mercury fulminate in the sample. The $KClO_3$ is found by subtracting the % of mercury fulminate + % moisture from 100%.

Analysis of Compositions Containing Nitrocompounds. Trinitrotoluene, tetryl or picric acid can be identified by melting point test, TNT. melting at about 79°–80° C., tetryl at about 128° C., and picric acid at about 120°–122° C. They may be extracted from the mixture by means of ethyl ether, in which mercury fulminate is only very slightly soluble, and the determination of $KClO_3$ and mercury fulminate then made as described in the preceding paragraph.

If the main charge is an organic nitrate such as nitrated vegetable ivory, nitrostarch, etc., such material will be left with the mercury fulminate in the insoluble residue after extraction with water. The mercury fulminate is then extracted by means of a hot 20% solution of sodium thiosulphate, leaving the organic nitrate in the Gooch crucible. These materials in the detonating composition can be readily identified by microscopic examination.

In detonators where TNT. or tetryl compose the main portion of the charge, a small amount of lead azide, with or without mercury fulminate may be used as a priming charge for the purpose of initiating the detonation of the nitrocompound. It should be identified in the top portion of the charge, next to the reinforcing cap, and will in all probability be present if mercury fulminate is not found. It is practically insoluble in water and in ether, and will be left in the insoluble residue. If present, fulminate is destroyed by treating the residue, in a flask, with 25 cc. of KOH solution. This converts the lead azide to potassium azide, KN_3. A slight excess of H_2SO_4 is added and the mixture distilled, the distillate, containing HN_3, being collected in water. Enough NaOH is added to the distillate to give an alkaline reaction with litmus, then a little $Pb(NO_3)_2$, when lead azide, PbN_6 will be regenerated as a white precipitate, which may be filtered off, washed with water, then with alcohol, dried in the air, and tested by striking a small portion with a hammer.

Primers

Variations in Composition. Many varieties of composition are used in primers for small arms ammunition, and for other military purposes. The composition must be ignited by the impact of the firing pin, and must give a flame of sufficient intensity and duration to ensure proper ignition of the propellant or of the detonator, depending on the purpose for which the primer is employed. As primers are used with various kinds and granulation of explosives, a priming composition suitable for one purpose is unsuited for another; hence there are many types of priming compositions, a few of which are indicated in the following table:

TYPES OF PRIMER COMPOSITIONS
APPROXIMATE COMPOSITION (PER CENT)

Ingredients	No. 1	No. 2	No. 3	No. 4	No. 5	No. 6	No. 7	No. 8	No. 9	No. 10
Mercury fulminate	31	25	11	..	..	..	..	..	..	28
Potassium chlorate	38	38	53	60	50	51	53	53	47	14
Sulphur	..	..	..	7	3	9	..	..	22	..
Powdered glass	..	..	..	..	..	12	..	..	..	35
Lead sulphocyanate	..	..	..	..	..	..	25	25	..	..
Copper sulphocyanate	..	..	..	3	..	..	..	..	..	..
Barium nitrate	..	6	..	..	..	..	..	..	..	..
TNT	..	..	..	..	..	..	5	..	..	..
Tetryl	..	..	..	..	..	..	..	3	..	..
Antimony sulphide	31	31	36	30	44	26	17	17	31	21
Lead oxide (PbO)	..	..	..	..	..	..	..	2	..	..
Shellac	..	..	..	..	..	2	..	..	..	2
Black powder (meal)	..	..	..	..	3	..	..	..	..	..

In addition to these ingredients most priming compositions are mixed with small amounts of some binding material dissolved in water or alcohol, such as gum arabic, gum tragacanth, glue, shellac, etc. These traces of binding materials are usually disregarded in the analysis of the compositions.

Preparation of Sample. If the caps contain anvils, these must first be carefully removed, as well as any covering of tin foil or paper. The primer composition is then carefully removed from a number of primers and weighed to determine the average charge. It is then carefully crushed, a little at a time, and the sample well mixed. If necessary, it may be removed from the caps by the aid of water or alcohol and the latter removed by evaporation before weighing.

Qualitative Examination. The following special tests may be made use of in connection with a qualitative analysis of the mixture:

A small amount is burned between two watch glasses, the formation of a mirror indicating mercury, antimony, copper or lead. The mercury mirror is readily volatile on gentle ignition.

Extract a portion of the mixture with ether, then with water, then with $Na_2S_2O_3$ solution, then with aqua regia, retaining each of these solutions.

TNT. or tetryl may be present in the ether solution and are identified by m.p. or color test, TNT. giving a deep red color with acetone and KOH. Sulphur is detected by burning a portion of the ether-soluble material and noting odor of SO_2.

The water extract is tested for $KClO_3$ by adding H_2SO_4, boiling, and noting odor of chlorine. A portion is treated with HCl and $FeCl_3$, a red color indicating thiocyanate. The usual $FeSO_4$ ring test is made for nitrates. A white precipitate with H_2SO_4 indicates Ba or Pb.

The aqua regia solution is diluted and tested with H_2S for antimony, lead, and copper. If the precipitate is not orange-red, lead or copper are indicated. Dissolve in HNO_3, neutralize with NH_4OH; a blue solution indicates copper, while lead is detected by the formation of a white precipitate with H_2SO_4.

Any material insoluble in aqua regia may be powdered glass or other abrasive material.

Quantitative Analysis. The method of analysis will depend entirely upon the ingredients indicated by qualitative tests. In general, a separation is best effected by successive extractions with ether, water, $Na_2S_2O_3$ solution (to remove fulminate), dilute or concentrated HCl, and aqua regia. The small amount of mercury fulminate present in the water extract may be determined by precipitation with H_2S or by adding 10–15 cc. of thiosulphate solution and a few drops of methyl orange and titrating with $N/10$ HCl or H_2SO_4 (see page 513 of Vol. I). Other materials in the water and acid solutions are determined by the usual analytical methods.

Nitrocellulose

General. The term nitrocellulose, or more correctly cellulose nitrate, applies to any nitration product of cellulose, ranging from products containing in the neighborhood of 10–11% N, which are used in the preparation of lacquers and other commercial products, to military guncotton with over 13% N. All of these products are undoubtedly mixtures of the various nitrates of cellulose, as indicated by the fact that there is always some material with low nitrogen content, soluble in ether-alcohol, in high nitrogen guncotton, and some insoluble material in the lower nitrated commercial products. It can usually be shown without great difficulty that any nitrated cotton is a mixture of various nitrates of cellulose.

The products of military importance are the insoluble guncotton of high N-content, and the so-called "pyro" or pyrocellulose, soluble in ether-alcohol and of about 12.60% N-content. In testing these products, the characteristics of most importance are content of nitrogen, solubility in ether-alcohol, and stability. Other determinations generally made are solubility in acetone and ash.

Preparation of Sample. If the sample contains a large excess of water, it is enclosed in a clean cloth and the excess water removed by means of a press or wringer. The pressed sample is then rubbed up in the cloth (not with the bare hand) until lumps are removed, then spread on clean paper trays in an air bath at about 35°–40° C. until "air-dry."

Samples for stability tests and nitrogen determination are treated as noted below, the air-dry sample being suitable for determining solubility and ash.

Nitrogen. About 1 to 1.05 g. of the air-dry sample is roughly weighed in a tared weighing bottle, dried at 95°–100° C. for $1\frac{1}{2}$ hours, cooled in a desiccator and accurately weighed. It is then transferred to the generating bulb of a nitrometer (Du Pont modification; see p. 354) using a total of 20 cc. of 95–96% c.p. H_2SO_4. The sample must be dissolved in the acid either in the weighing bottle or in the cup of the generator, before it is drawn into the generating bulb, and both the weighing bottle and the cup of the generator must be thoroughly washed out with the 20 cc. of H_2SO_4, so that none of the sample is lost. The determination in the nitrometer is completed in the usual manner (p. 354), the result being expressed as % N in the dried sample of nitrocellulose.

Solubility in Ether-Alcohol. (a) *Guncotton:* The amount of ether-alcohol soluble material in guncotton being usually not more than 10–12%, the determination may be made by evaporating a clear solution. Two grams of air-dry sample is placed in a clean dry cork-stoppered 250 cc. cylinder, 67 cc. of 95% ethyl alcohol added to thoroughly wet the guncotton, then 133 cc. of ethyl ether (U.S.P. grade, 96%), added and the mixture well shaken. If the mixture of 2 parts ether and 1 part alcohol be added at once to the sample, a gummy mass may result which dissolves with great difficulty, especially if the solubility is unusually high.

The cylinder is now allowed to stand at a constant temperature of usually 20° C. (15.5° C. is sometimes specified). The solubility of nitrocellulose *increases* as the temperature is *decreased*, hence a constant temperature of digestion is important. During the digestion, which requires at least 1 hour, the cylinder must be thoroughly shaken at 5-minute intervals. The cylinder is now allowed to stand for at least 4 hours, until the insoluble portion of the sample has completely settled and the supernatant liquid is perfectly clear.

50 cc. of the clear solution is now drawn off with a pipette, care being taken not to disturb the settled pulp, and evaporated in a weighed evaporating dish on a steam bath, avoiding loss from violent boiling of the ether. When 25–30% of the solution has been evaporated, 10 cc. of distilled water is added slowly and the evaporation continued to dryness. The effect of the water is to leave the residue in a white, brittle or powdery condition, rather than a tough film which would lose its solvent with difficulty.

The dish is finally placed in an oven at 95–100° C. for $\frac{1}{2}$ hour, cooled in a desiccator, and weighed. The weight of the residue, corrected for the residue in the 50 cc. of ether-alcohol and 10 cc. H_2O used, represents the soluble nitrocellulose in 0.5 g. of the guncotton.

(b) *Pyrocellulose:* The solubility of pyrocellulose may be determined in the manner described for guncotton, but owing to the much larger amount of soluble material present, the evaporation of the residue to constant weight without decomposition involves considerable difficulty. Sufficient water must be added to precipitate the soluble nitrocellulose from solution in a stringy or fibrous condition.

The determination is usually conducted by either the volumetric method or the filtration method.

In the volumetric method, one gram of the air-dry sample is covered with 100 cc. of 95% ethyl alcohol and allowed to stand at least 15 minutes with frequent stirring, 200 cc. of ethyl ether is then added with stirring and the agitation continued until solution is complete. The solution is now allowed to stand at least 4 hours with frequent stirring, during at least 1 hour of which time it is to be kept at a temperature of 15.5° C. It is then transferred to a "solubility tube" and allowed to stand for at least 16 hours, in order that the insoluble material may settle completely. The solubility tubes are glass tubes about 30.6 inches long×1.3 inches inside diameter, tapering at a point 6 inches from the bottom to a constricted portion about 3 inches long and about .375 inch inside diameter. This narrow bottom portion is graduated to read directly the percentage of insoluble material, the value of the graduations having been first ascertained by comparison with results obtained by the filtration method described below. The tubes are made of heavy glass and provided with vented ground glass stoppers. They hold 300 cc. when filled to about 8 inches below the top.

In the filtration method, the solution is prepared and settled in a solubility tube as described above, and the clear liquid removed as completely as possible by means of a narrow siphon tube of glass. Fresh alcohol and ether are then added as before, the tube shaken and allowed to stand again for 16 hours, when the process may be repeated several times, depending on the amount of insoluble material present. After the last decantation, the residue is washed from the tube to a beaker, using as small a quantity of ether-alcohol as possible, and the mixture filtered through a filtering tube consisting of a $1''\times6''$ test tube with its lower end drawn out to a taper terminating in a hole about $\frac{1}{8}''$ diameter. In the lower end of this tube is a small plug of previously ignited asbestos. The filtration is facilitated if the greater part of the asbestos is mixed with the insoluble matter and solvent in the beaker, the mixture well stirred and quickly poured into the filtering tube on top of a small plug of asbestos. In this manner, the insoluble matter becomes mixed with the asbestos and the formation of a gelatinous, impenetrable mat in the tube is avoided. After filtering, the tube is washed with fresh ether-alcohol, dried at 40°–45° C. and finally for 1 hour at 100° C., then cooled in a desiccator and weighed. All combustible matter is then removed by careful ignition, and the tube again weighed, the loss of weight being the total insoluble material in the 1-gram sample.

Solubility in Acetone. A 1-gram sample of air-dry pyrocellulose is treated with about 200 cc. of acetone with frequent stirring until all gelatinous matter has dissolved. The solution is transferred to a solubility tube (described above), the volume made up to about 300 cc. with fresh acetone, well shaken, and allowed to settle for at least 16 hours. The graduations on the tube having been checked by gravimetric determinations, the percentage of residue insoluble in acetone may be read direct, or the filtration method described above may be applied.

Ash. One gram of air-dry sample is weighed in a tared crucible, moistened with 10–15 drops of concentrated nitric acid, and digested for 2–3 hours on a steam bath until converted to a gummy mass. The crucible is then heated carefully over a Bunsen burner until the mass is completely charred, then at a red heat until its weight is constant. The residue is the ash of the sample.

Stability Test: Heat test with Potassium Iodide Starch Paper. The "heat test" or KI test, as it is commonly designated, is the test most commonly employed for determining the stability or degree of purification of nitrocellulose, whether guncotton or pyrocellulose. This test, also referred to as the Abel test, depends on the action of oxides of nitrogen liberated by the nitrocellulose under the influence of heat, the gases in contact with the KI-starch paper liberating iodine which colors the starch.

The sample is dried with great care to avoid contamination, in a clean paper tray, at 35° to 43° C., until its moisture is reduced to the amount which will give the minimum heat test, usually 1.5 to 2%. The proper amount of moisture is determined as follows: During the progress of the drying, the sample on the tray is "rubbed up" from time to time, using a piece of clean tissue paper spread over the back of the hand. When the sample begins to adhere to the paper, due to static electricity, a sample of 1.3 g. is weighed into a standard test tube. These tubes are $5\frac{1}{2}$ inches long, not less than $\frac{1}{2}$ inch inside diameter and not more than $\frac{5}{8}''$ outside diameter, made of glass about 3/64 inch (1.2 mm.) thick. As soon as the first sample is weighed, the tray is replaced in the drying oven for 2–5 minutes, a second sample weighed, and this process repeated until a series of 5 samples have been taken, the last sample being completely dry. This series of samples, if properly taken, will cover the range of moisture content giving the minimum heat test. If the sample in the tray appears to have become too dry during the time the weighings are being made, it may be placed in a moist atmosphere for not more than 2 hours; the entire time of drying and making the test must not exceed 8 hours.

The tubes containing the samples are fitted with clean, fresh cork stoppers through which pass a piece of glass rod into the end of which is fused a small piece of platinum wire bent into a hook. The wire is heated in a flame to clean it, a piece of the standard KI starch test paper, $1'' \times \frac{3}{8}''$, attached, taking care that neither wire nor paper are touched with the fingers, and the paper moistened on its upper portion by touching it with a glass rod dipped in a solution of equal volumes of pure glycerin and water. The stoppers are then inserted in the tubes and the tubes placed in a constant temperature water bath, so that they are immersed to a depth of 2.25 inches. The time of placing in the bath and the time of the appearance of the first faint yellowish discoloration of the test paper are noted. The minimum test given by the 5 samples is taken as the result of the test. The discoloration appears at the lower edge of the moist portion of the paper. The temperature of the heat test bath is 65.5° C. (150° F.) for pyrocellulose, and usually 76.5° C. (170° F.) for guncotton. Pyro is usually required to stand a test of 35 minutes, and guncotton 10 minutes.

A standard test paper is absolutely essential, and is prepared as follows:[1]

The paper used in preparing the test paper is Schleicher and Schüll's filter paper 597. This is cut in strips about 6 by 24 inches, and after being washed by immersing each strip is distilled water for a short time is hung up to dry overnight. The cords on which the paper is hung are clean and the room is free from fumes. The washed and dried paper is dipped in a solution prepared as follows:

The best quality of potassium iodide obtainable is recrystallized three times from hot absolute alcohol, dried, and 1 gram dissolved in 8 ounces of distilled water. Cornstarch is well washed by decantation with distilled water, dried at a low temperature, 3 grams rubbed into a paste with a little cold water, and poured into 8 ounces of boiling water in a flask. After being boiled gently for 10 minutes, the starch solution is cooled and mixed with the potassium iodide solution in a glass trough.

[1] Storm, C. G., Proc. 7th Inter. Congress Appl. Chem., 1909; J. Ind. & Eng. Chem., vol. 1, 1909, page 802.

Each strip of filter paper is immersed in the above-mentioned mixture for about 10 seconds and is then hung over a clean cord to dry. The dipping is done in a dim light and the paper left overnight to dry in a perfectly dark room. Every precaution is taken to insure freedom from contamination in preparing the materials and from laboratory fumes that might cause decomposition. When dry the paper is cut into pieces about ⅜ by 1 inch and is preserved in the dark in tight glass-stoppered bottles, the edges of the large strips being first trimmed off about one fourth inch to remove portions that are sometimes slightly discolored. When properly prepared the finished paper is perfectly white, any discoloration indicating decomposition due to contamination.

Stability Test at 135° C. In addition to the KI starch test, pyrocellulose is usually required to stand a test at 135° C., made as follows:

The sample is completely dried at 42° C., and 2.5 grams placed in each of 2 heavy glass tubes, 290 mm. long, 18 mm. outside diameter and 15 mm. inside diameter, closed with a cork stopper through which passes a hole 4 mm. in diameter. A strip of litmus paper or standard normal methyl violet paper, 70 mm. long and 20 mm. wide is placed in each tube, its lower edge 25 mm. above the sample, which is pressed down to occupy a depth of 2 inches, the walls of the tube being wiped clean with a roll of paper. The tubes are then heated in a constant temperature bath at 134° to 135° C., all but about 6–7 mm. of the tube being immersed in the bath. They are partially withdrawn for examination of the test papers every 5 minutes after the first 20 minutes of heating, and replaced at once. The time required for reddening of the litmus paper or for turning the methyl violet paper to a salmon pink color is noted as the time of the test. A minimum test of 30 minutes is required with the methyl violet paper, and heating is then continued for a total of 5 hours, during which time there should be no explosion.

The standard normal methyl violet paper is prepared as follows:

Preparation of Methyl Violet Test Paper. A solution is prepared containing the following ingredients: pure rosaniline acetate prepared from 0.2500 g. basic rosaniline, .1680 g. methyl violet (crystal violet), 4 cc. c.p. glycerin, 30 cc. water, and sufficient pure 95% ethyl alcohol to make up to 100 cc. This solution is placed in the angle of an inclined deep rectangular glass tray, and large sheets of Schleicher & Schüll filter paper (No. 597) cut in four strips are dipped in it. In dipping, the strip is held by one end and dipped to within ¼″ of this end, withdrawing it slowly up the side of the tray so as to remove surplus solution. The strip is then held horizontally and waved to and fro so as to prevent the solution from running and collecting in spots. As soon as the alcohol has evaporated the strip is suspended vertically to dry, and when dry is cut in strips 20×70 mm. These strips are bottled and kept for use in the 135° test.

SMOKELESS POWDER

Nitrocellulose Powders

At the present time the smokeless powder used by all nations is composed of either colloided nitrocellulose alone or a mixture of colloided nitrocellulose and nitroglycerin. All cannon powder used in this country is of the nitrocellulose type, small-arms powders being of both types. The form and size of the grains are of great variety, depending on the arm in which the propellant is to be employed.

Physical tests made in connection with the examination of smokeless powder include the compression test, determinations of average measurements of the grains, specific gravity, gravimetric density, number of grains per pound, and calculation of burning surface per pound.

Chemical tests include determinations of moisture and volatile solvent, diphenylamine used as stabilizer, ash, material insoluble in ether-alcohol and in acetone, and sometimes nitrogen content.

Stability tests include the 135° C. test, the 115° C. test, and the "Surveillance test."

Moisture and Volatiles. A sample of the powder weighing approximately 1 gram, in the form of thin shavings cut from at least 10 grains, or of whole grains if the powder is too small to cut conveniently, is placed in a clean, dried and weighed 250 cc. beaker, 50 cc. of redistilled 95% (by volume) alcohol, and 100 cc. redistilled ethyl ether added and the beaker allowed to stand under a cover-jar with occasional stirring, until the powder is completely dissolved. This usually requires from 1 to 2 days. When all gelatinous particles of the powder have dissolved, the beaker is heated on the steam bath to evaporate a part of the ether, before precipitation of the nitrocellulose with water. The amount of ether to be evaporated is important, since it largely determines the character of the nitrocellulose precipitate. The presence of too much ether causes a fine sandy precipitate; too little causes a gummy, gelatinous precipitate. A fine, flaky, or fibrous precipitate is desirable. The proper amount of evaporation can be best determined by practice; usually the solution may be evaporated to about $\frac{2}{3}$ its original volume before precipitating. When the proper volume is obtained, 50 cc. of water is added from a graduate, with continual stirring, in 5 cc. portions. If a thick gummy precipitate forms, add a little ether until it becomes flaky; then add the remainder of the 50 cc. of water. The heating is continued with stirring, until most of the ether has evaporated, and the beaker is then left on the bath until the precipitate is just dry. It is then placed in the 100° C. oven for 1 hour, cooled in a desiccator, and weighed as rapidly as possible. To facilitate weighing the weights should be placed on the balance pan before the beaker is removed from the desiccator, so that the exact weight can be adjusted quickly. If more than 10 seconds are consumed in this weighing, the error caused by absorption of moisture from the air is an appreciable one. In any event a check weighing should be made after an additional 30 minutes drying at 100° C.

The final weight of nitrocellulose precipitate subtracted from the weight of the original sample represents the weight of moisture and volatile solvent, and is calculated as per cent of the original sample. If the powder contains diphenylamine, this result is corrected by subtracting from it one fourth of the total diphenylamine content, it having been ascertained by actual trial that approximately this proportion of the diphenylamine is volatilized during the evaporation.

Moisture. An approximation to the actual moisture content of the powder can be obtained by drying a sample of not less than 5 whole grains and not less than 20 grams for 6 hours at 100° C., cooling in a desiccator and weighing, the loss of weight being regarded as equal to the hygroscopic moisture in the powder.

Diphenylamine. The content of diphenylamine used as a stabilizer in smokeless powder is most conveniently and rapidly determined by the "nitration method" as follows:

5 grams of the powder in small grains or slices is treated with 30 cc. of concentrated HNO_3 in a 250 cc. beaker, covered with a watch glass and heated on the steam bath until the powder has been completely decomposed. The solution is then cooled and added to 100 cc. of cold distilled H_2O in a second beaker, stirring vigorously, the first beaker being washed out completely into the second, using additional water. This mixture is now heated on the steam bath until the flocculent precipitate has settled and the liquid has a clear yellow color. It is then cooled, filtered through a weighed Gooch crucible, the precipitate dried at 100° C. and weighed. The weighed precipitate is now dissolved by extracting with acetone, the crucible dried and weighed again, the loss of weight being the nitrodiphenylamine produced by action of the HNO_3 on the diphenylamine. This nitrodiphenylamine is a mixture of nitroproducts, and the empirical factor 0.40576 has been determined for converting it to its equivalent in diphenylamine.

Ash. The ash is determined in the manner described for nitrocellulose (p. 1401), the sample being in the form of slices or small grains, and the digestion with HNO_3 continued until decomposition is practically complete, before heating over a flame.

Solubility in Ether-alcohol. One gram of the sample in slices or small grains is dissolved in 150 cc. of ether-alcohol (2 : 1) in the same manner as for the determination of moisture and volatiles, and transferred to a standard solubility tube (p. 1401), washing it in completely with fresh ether-alcohol so as to bring the total volume to 300 cc. The insoluble material is determined as in pyrocellulose (p. 1401).

Solubility in Acetone. This determination is made in the same manner as the solubility in ether-alcohol, described above.

Stability Test at 135° C. This test is made on duplicate samples in the same manner as described for pyrocellulose (p. 1403). The samples weigh 2.5 grams and are in as nearly whole grains as is consistent with this weight of sample, large grains being turned down on a lathe to fit the standard tubes. The samples are required to stand heating at 134°–135° C. for 5 hours without explosion and must not turn the normal methyl violet paper to salmon pink color in less than one hour.

Stability Test at 115° C. This test is also known as the Ordnance Department 115° test, or the Sy test. Five samples each consisting of not more than 10 grams and not less than 2 whole grains of the powder are weighed on watch glasses and heated at a temperature of 115° ± 0.5° C. for 8 hours daily for 6 days, the oven being brought each day to the proper temperature before the samples are inserted, the samples being allowed to stand at room conditions overnight. At the end of the sixth day's heating, the samples are cooled in a desiccator and weighed. The total loss of weight is regarded as an index of the stability, and must not exceed a specified limit for each particular size of grain.

"**Surveillance Test**" at 65.5° C. Three samples of approximately 45 grams of powder in whole grains, or, in the case of very large grains, 5 whole grains, are placed in 8-ounce wide-mouth glass stoppered bottles, the stoppers having been previously ground so as to fit tightly. These bottles are then heated in a constant temperature magazine at 65.5° ± 2° C. They are observed several times daily and the time noted when visible fumes of oxides of nitrogen appear in any bottle. The number of days which powder is required to stand this test depends on the web thickness of the grain, and varies from 70 to 140 days. The test is therefore not a laboratory test, but one which more nearly approaches service conditions. It is of great value as an indication of the possible "stability life" of the powder in service.

Nitrogen. The determination of nitrogen in smokeless powder is not usually necessary, in as much as the powder is usually made from nitrocellulose of known nitrogen content, but when desired the determination is made as follows:

An average sample of about 5 grams of the powder in slices or small grains is dissolved in acetone (100 cc. to each 1 g. of sample). When the sample is dissolved, the solution is added drop by drop, preferably from a burette, to 200 cc. of hot water in a beaker, the beaker being immersed in boiling water so as to maintain its contents at about 90° C. During this addition the hot water is continually stirred with a glass rod, so that the precipitated nitrocellulose forms stringy masses which wrap about the rod. Small accumulations of the precipitate are transferred frequently from the rod to another beaker of hot water to prevent the formation of a colloided mass. When 2 g. or more of the precipitate has been collected and the acetone has been volatilized by the hot water, it is removed from the beaker and dried at 35°–40° C. About 1 g. of this dry precipitate is placed in a tared weighing bottle, dried 1 hour at 100° C., weighed, and transferred to the cup of the nitrometer with sulphuric acid. Part of the acid should be added to the precipitate in the weighing bottle before transferring to the nitrometer in order to avoid loss of the dry precipitate in handling. The determination of N is then completed as in the case of nitrocellulose (page 1400). If the powder contains diphenylamine, a correction is necessary for the amount of diphenylamine retained by the precipitated nitrocellulose. This has been found to be an added correction of 0.15% N in the case of powders containing the usual amount of 0.4% diphenylamine. This correction compensates for the nitrogen which becomes combined with the diphenylamine, converting it to nitrodiphenylamines.

Instead of correcting for the effect of the diphenylamine, the latter may be removed from the precipitated nitrocellulose, after air-drying and before final drying at 100° C., by extraction with pure anhydrous ether. Results are quite accurate if the determination is conducted with proper precaution.

Nitroglycerin Smokeless Powders

Powders of this type are composed mainly of nitrocellulose and nitroglycerin and may contain other organic or inorganic substances, such as vaseline, nitro-substitution compounds, substituted ureas or other flame-reducing or surface-hardening agents, diphenylamine, metallic nitrates, carbonates, etc. The nitrocellulose may be either high-nitration guncotton insoluble in ether-alcohol, as in British cordite, or a low-nitration product soluble in nitroglycerin, as in ballistite, or may be a mixture of the two varieties.

The method of analysis usually employed consists of (1) an extraction of the nitroglycerin, nitrosubstitution compounds, vaseline, and other ether-soluble materials by means of anhydrous ether; (2) an extraction of the water-soluble materials; (3) determination of soluble and insoluble nitro-celluloses by separation with ether-alcohol (2 : 1).

The extraction with ether is usually made in a Soxhlet apparatus, using about 20 grams of the powder in slices or small grains, in a paper extraction thimble. About 4 hours is usually required for complete extraction. The ether extract is evaporated to dryness in a tared glass dish under a bell-jar evaporator (page 1376), and the ether-soluble residue weighed. To determine whether it contains other substances than nitroglycerin, it may be poured in small portions at a time into about 20 cc. of strong nitric acid (40° Be) heated on a steam bath. The oxidizing action of the nitric acid destroys the nitroglycerin, and the mixture is then poured into 50–100 cc. of water. Any vaseline or similar substances separate, together with any nitrosubstitution compounds in their original condition or more completely nitrated, diphenylamine in the form of a nitroderivative, etc.

These materials may be separated with more or less completeness by fractional crystallization from ether or other solvent. The exact method to be followed depends on the nature of the materials present.

The residue insoluble in ether is dried and weighed, and then transferred to an Erlenmeyer flask and digested in warm water until any water-soluble materials present have been dissolved. The mixture is filtered, the residue washed with hot water, dried and weighed. The filtrate containing the water-soluble ingredients is examined by the usual analytical methods for inorganic ingredients.

The nitrocellulose insoluble in water is tested for nitrogen content, solubility in ether-alcohol and solubility in acetone, by the methods already described.

Typical Compositions of Commonly Used Explosives

Black Blasting Powder

Sodium nitrate....................73
Charcoal..........................16
Sulphur...........................11 (Bu. of Mines, Bull. No. 80, p. 19.)

Black Military Powder

Potassium nitrate.................75
Charcoal..........................15
Sulphur...........................10

Typical Dynamite formulas—40% grades (Bu. Mines, Bull. No. 80, p. 21).

	Nitro-glycerin	Nitro-Substitution Com.	Ammo-nium Nitrate	Sodium Nitrate	Nitro-cellu-luse	Wood Pulp	Calcium Carbonate
40% straight Nitroglycerin Dynamite	40	......	......	44	......	15	1
"40%-strength" Ammonia Dynamite	22	......	20	42	......	15	1
"40%-strength" Gelatin Dynamite..	33	......	......	52	1	13	1
"40%-strength" Low-freezing Dynamite...........................	30	10	......	44	......	15	1
"40%-strength" Low-freezing Ammonia Dynamite.............	17	4	20	45	......	13	1

Granulated Nitroglycerin Powder ("Judson Powder",

Nitroglycerin....................................5   10
Combustible material†........................35   26
Sodium nitrate..............................60   64

Coal Mining Powders. (Permissible Type)

	I	II	III	IV	V	VI
Nitroglycerin...........................	25	15	10	......	......	10
TNT....................................	......	5	......	5	......	......
Ammonium nitrate.......................	......	......	79	90	94	70
Sodium nitrate.........................	34	35	......	......	......	......
Sodium chloride........................	......	......	......	......	......	9
Wood pulp..............................	15	12	10	......	......	10
Flour..................................	25	17	......	5	......	......
Aluminum powder........................	......	......	......	......	3	......
Charcoal...............................	......	......	......	......	3	......
Calcium carbonate......................	1	1	......	......	......	1
Zinc oxide.............................	......	......	1	......	......	......
Magnesium sulphate, cryst..............	......	15	......	......	......	......

* Sometimes contains also flour, cornmeal, sulphur, etc.
† Composed of sulphur, coal, and rosin.

www.ingramcontent.com/pod-product-compliance
Lightning Source LLC
Chambersburg PA
CBHW021602210326
41599CB00010B/568